Foundations of a Pure Cost Theory

Heinrich von Stackelberg

Foundations of a Pure Cost Theory

Translated by Damien Bazin
(Scientific Director),
Lynn Urch and Rowland Hill

Springer

Heinrich von Stackelberg (1905–1946)
Professor of Economics
University of Bonn
Bonn
Germany

Translators
Damien Bazin
Professor of Economics
University of Nice Sophia-Antipolis
UFR Higher Institute of
Economics and Management (ISEM)
National Centre for Scientific Research
GREDEG UMR7321 UNS-CNRS
rue Albert Einstein
Valbonne – France

Lynn Urch
Gloucester
United Kingdom

Rowland Hill
Interlink
Le Château d'Olonne
France

Von Stackelberg "Foundations of a Pure Cost Theory"
Translated by Damien Bazin (Scientific Director),
Lynn Urch and Rowland Hill

Translation from the German language edition: "Grundlagen einer reinen Kostentheorie", by Heinrich von Stackelberg © Springer-Verlag Vienna New York 1932, all rights reserved.

ISBN 978-3-642-34536-4 ISBN 978-3-642-34537-1 (eBook)
DOI 10.1007/978-3-642-34537-1
Springer Heidelberg New York Dordrecht London

Library of Congress Control Number: 2013939979

Printed on acid-free paper

Springer is part of Springer Science+Business Media (www.springer.com)

I would like to express my very sincere thanks to my esteemed supervisor, Professor Erwin von Beckerath of Cologne, for the valuable support he has shown my work in every respect. I would like to warmly thank Professor Hamburger of Cologne, for his valuable advice about formulating the mathematical representations. Similarly, I owe a debt of gratitude to Dr. Morgenstern of Vienna, and the publishing house, Julius Springer Verlag, Vienna, for their kind cooperation which has enabled us to publish this work.
Rome, October 1932.
Heinrich von Stackelberg

This Book is Dedicated To My Mother
Baroness Luisa Von Stackelberg
Nee De Vedia

To my daughter, Lily, with love

To my husband, Mike, and daughter, Sophie, with love

Foreword

The translation of valuable work by a noted contributor to an academic field evidently needs no justification. But for the works of Heinrich Freiherr von Stackelberg, there are special reasons for commentary.

Germany has long been recognized as the source of significant research and teaching in economics. Indeed, in the nineteenth century for a number of noted economists in the United States and Great Britain, periods of study at German universities clearly augmented their qualifications. Interestingly, attention was largely directed to a group of investigators, the German Historical School, who largely eschewed not only mathematical analysis but even formal theory of the sort that had emerged in Great Britain in the wake of Adam Smith and David Ricardo's contributions.

In the German-speaking world, formal theory apparently was left to Austrian researchers, who made significant contributions to capital theory, the use of marginal analysis in the theory of value, and other pertinent areas. But even in Austria where the work of noted scholars, such as Carl Menger, Friedrich von Wieser, and Eugen von Böhm-Bawerk, provided significant theoretical analysis, there was little reliance on mathematical tools in economics. Indeed, the most fundamental Austro–Hungarian breakthrough, the introduction into economics of the theory of games, came soon before World War II and was contributed by the partnership of John von Neumann and Oskar Morgenstern largely *after* they arrived in the United States.

Still it is difficult to believe that Germany failed to produce *any* masters of the mathematical techniques already being used in economics and the growing body of economic analysis. After all, Germany was the source of revolutionary breakthroughs in the field of pure and applied mathematics, so it surely is implausible that this surge of attention to mathematical analysis would have left the economics discipline unscathed.

Indeed, the author of this book helped to lead Germany's transition to incorporation of illuminating mathematical analysis into economic issues. Coming at a time when economic theory dealing with the behavior of firms and markets was breaking through the circumscription of the Marshallian model, Stackelberg's contributions could have been expected to occupy a fuller place in the literature—had war and totalitarian regimes not attracted attention to other matters. Indeed, it is startling to recognize the inadequacy of the attention this book and Stackelberg's other work have received.

Granted, Stackelberg's writings have been discussed by many noted economists, including luminaries such as Nicholas Kaldor and my former colleague Fritz Machlup. And an English translation (1952; by my friend Alan Peacock, incidentally) of his *Grundlagen der Theoretischen Volkswirtschaftslehre* was published 60 years ago. Still, the unfortunate fact remains that Stackelberg's work is all too often overlooked—for instance, in the various tomes on the history of economic ideas. This may not be surprising in an elementary textbook, such as Eric Roll's much utilized volume,[1] but we would expect the situation to be very different when we turn to Schumpeter's great work *History of Economic Analysis*,[2] which is widely known not merely for its size but also for leaving very few stones unturned. Moreover, it is evident that its author was fluent in German and strongly supported the use of mathematical tools in economic analysis. Yet here too, the reader will be disappointed. Stackelberg appears only in a handful of footnotes, generally as only one of several references to the various literature pertinent to Schumpeter's discussions in the main text.

The implications are clear. Stackelberg, an analyst writing in the spirit of his time, provided substantial insights that were in keeping with the new directions that economic theory was taking in the years soon after World War I in the United States and Great Britain. And as I have emphasized here, the bulk of the literature on the history of economic ideas has given far too little attention to his work.

The present volume may help us to address this latter shortcoming. Clearly a labor of love, it offers the members of our discipline a new opportunity to gain a better understanding of developments in economics near the time this book was written. Indeed, we should be grateful to the translator-editors of this book for their persistent efforts to bring Stackelberg's significant and valuable contributions into the mainstream economics literature.

Princeton, NJ William J. Baumol

---

[1] Roll, E. (1953), *A History of Economic Thought*, Englewood Cliffs, New Jersey: Prentice Hall.

[2] Schumpeter, J.A. (1954), *History of Economic Analysis*, New York: Oxford University Press.

Praise for the book

Today's economics students only learn about Stackelberg in two contexts, the leader-follower equilibrium and duopoly warfare. And within these two contexts the focus remains invariably on modern game-theoretic formulations following Spence (1977) and Dixit (1980); any mention of Stackelberg is confined to the footnotes. Therefore this new translation of an earlier book, which constituted Stackelberg's doctoral dissertation—the better-known work was his second, or Habilitation dissertation—is to be welcomed.

On reading the book I was truly impressed by Stackelberg's analysis of costs of production. His emphasis on the time aspect, as exemplified by his use of the term "velocity of production," which serves as a constant reminder that what we now casually call the volume or the quantity of production is actually a rate per unit time, deserves to be better known and remembered. His formulation of cost functions of multiproduct firms is an early precursor of the very sophisticated and fruitful line of analysis by William Baumol and others (Baumol 1982, Baumol, Panzar and Willig 1982). And he has some interesting early analysis of vertically related firms and transfer pricing.

The economics profession, and especially students of the history of economic thought, owe the translators much gratitude for their work in making this book available in English and drawing renewed attention to it.

Princeton, NJ
April 2013

Avinash Dixit

References

Baumol, W. J., (1982). Contestable markets: an uprising in the theory of industry structure. *American Economic Review*, *72*, *No. 1*, March, 1–15.

Baumol, W. J., Panzar, J. C. and Willig, R. W. (1982). *Contestable markets and the theory of industry structure*. San Diego, CA: Harcourt Brace Jovanovich.

Dixit, A., (1980). The role of investment in entry-deterrence. *Economic Journal*, *90*, *No. 1*, March, 95–106.

Spence, A. M., (1977). Entry, capacity, investment and oligopolistic pricing. *Bell Journal of Economics*, 8, No 2, Autumn, 534–544.

Comments on the Translation

We were not planning to start work on another of Baron Heinrich Freiherr von Stackelberg's books so soon after finishing our first. We were therefore surprised to be approached again by Springer, the prestigious publishing house, to translate *Grundlagen einer Reinen Kostentheorie* (1932) into English. In view of the success enjoyed by our first translation of *Marktform und Gleichgewicht* (1934)/*Market Structure and Equilibrium* (2011) among students, the university community and the academic world in general, we felt that translating another major work such as *Grundlagen einer Reinen Kostentheorie* (1932)/*Foundations of a Pure Cost Theory* (2013) would be of great benefit to the scientific community as a whole. Moreover, enthusiasm for the book made us realize the importance of continuing to provide as comprehensive an overview as possible of a body of scholarly work completed by one of the founding fathers of the Freiburg school of economic thought,[3] work which despite engendering initial controversy for its highly revolutionary and paradigmatic tone, was unanimously accepted. It is interesting to note that Stackelberg's doctoral thesis for the University of Cologne, which was supervised by Erwin von Beckerath and presented on 16 July 1930, was entitled, *Grundlagen einer Reinen Kostentheorie* (*Foundations of a Pure Cost Theory*). This PhD dissertation established the reputation of its author, as Stackelberg immediately became recognized (or rather, restored to favour) by his peers. Strangely, this book is not his best known work, although four different versions of it exist. Indeed, according to Hans Möller, Stackelberg's pupil, there are four different and complementary versions of *Grundlagen*: the original dissertation text in the University doctoral examinations files in Cologne; the two articles in the "Zeitschrift für Nationalökonomie" Vol. III (1931/32), Part 1, Number 3, pp. 333–367 and Part 2,

[3] Stackelberg's work still seems so fresh; we want to encourage the reader to explore this in greater depth via Eucken's *Ordnungstheorie* because it presents a renewed and atypical vision of the way markets operate (see "*vollständiger Wettbewerb*") through the ordinal utility of (stable) prices. However, this vision is at odds with the traditional German Historical School (Fichte, List, Müller and von Savigny). Not only does Stackelberg give ordoliberalism pride of place in the economy, but Eucken, Sombart and Spiethoff do likewise.

Vol. 4, pp. 552–590, hence, 74 p. in total in Vienna; the official bound copy of the dissertation consisting of the articles described above, including a special Preface, Contents and Biography; and finally, the monograph, Vienna 1932 (131 p.) with the impressum, "*Erweiterter Sonderdruck aus Zeitschrift für Nationalökonomie*, Bd. III, H. 3 und 4" (Extended reprint taken from the "Zeitschrift für Nationalökonomie" (Journal of Economics), Vol. III, Numbers 3 and 4).[4] Although *Grundlagen* (*Foundations*) was widely distributed and discussed, the fact remains that since the 1950s, the literature[5] has given preference to citing *Marktform und Gleichgewicht*.[6] With this translation, we hope to redress the balance, albeit in a modest way.

Foundations of a Pure Cost Theory will enable readers to explore the author's central tenets and gain an understanding of his thought processes. His mathematical reasoning is very powerful and still has an influence on contemporary debate. This 1932 book deals with the concepts of production, cost theory and the structure of the market. The final chapter in particular reveals a link with the 1934 book through its examination of basic and fundamental economic principles as well as market structures (focusing on imperfections within the competition model); obviously this concept has been fully explained according to clearly defined levels of abstraction.

We learned much from the experience of translating the first book, *Marktform/ Market Structures*, and hope that we have succeeded in transcribing as faithfully as possible the quintessential nature of Stackelberg's fertile thinking as well as his great finesse.[7] As such, *Foundations* has prompted numerous discussions in the academic world (notably, Tinbergen 1933, Schneider 1933, Hicks 1935 etc.). *Market Structure and Equilibrium* has also opened up an extensive debate about the writings of Hick 1935, Lange 1935, Vleugels 1935, Weinberger 1935, Zeuthen 1935, Kaldor 1936, Leontief 1936, Heyward 1941 and Rothschild 1947, etc.[8]

This translation project, which has sometimes been difficult and often arduous, but always fascinating, would not have been possible without the hard work of Lynn Urch and Rowland Hill, who produced a very high quality translation and succeeded in finding just the right words, despite the obstacles they encountered.

In conclusion, we would like to draw the reader's attention to the inherent risks associated with translating any work. The very act of translation means that we are

---

[4] See Moller, H. (1992), Teil III (Part III), "Die abgedruckten Texte" (Reproduced Texts), Abteilung A (Section A), p. 3.

[5] See the famous debate between Fellner and Stackelberg.

[6] *Market Structure and Equilibrium* represents the (mature) work of a university academic, enabling the author to obtain his habilitation, presented on 4 January 1935 (*Habilitationsschrift*), and at the same time establishing him as a distinguished economist. This habilitation, which will become one of Stackelberg's finest works, will go on to be recorded by Kaldor in *Economica* and Leontief in the *Journal of Political Economy*.

[7] Stackelberg's scholarship clearly extends beyond the bounds of orthodox economics (he was a polyglot and intellectually very close to Haberler, Machlup, Morgenstern and Von Hayek on the one hand and Amoroso on the other).

[8] See Moller, H. (1992, p. 68) for the full list of references.

prone to a certain amount of subjectivity, and as a result, a certain amount of interpretation, although we have attempted to minimize this. As scientific director, I assume all responsibility for any failures, errors or shortcomings due to the quality of our translation. Nevertheless, I am keen to emphasize that we have been scrupulous in our efforts not to distort the author's words or intentions. Some lexical and syntactical choices have been made in order to facilitate the comprehension of the book, but without losing sight of the essence of the original text. The aim of this translation is to remain neutral and free of any controversy concerning its author (particularly in relation to the period of his youth between 1931 and 1937) and his essays on economic policy.[9]

Dr. Damien Bazin

---

[9] We refer the reader to the writings of Hans Möller (1949 and 1992) for a complete biography of Stackelberg. See also James Konow (1994) for a critique of Stackelberg's (corporatist) political theory.

Moller, H. (1949), *Heinrich Freiherr von Stackelberg und sein Beitrag für die Wirtschaftswissenschaft* (Heinrich Freiherr von Stackelberg and His Contribution to Economics), "Zeitschrift für die Gesamte Staatswissenschaft"(*Journal of Political Science*), Tübingen.

Moller, H. (1992), *Heinrich Freiherr von Stackelberg. Gesammelte wirtschaftswissenschaftliche Abhandlungen* (Heinrich Freiherr von Stackelberg. Collected Economic Reviews), vol. 1, edited by Norbert Kolten and Hans Möller, Regensburg, Transfer Verlag, 82 p.

Konow, J. (1994), "The Political Economy of Heinrich von Stackelberg", *Economic Inquiry*, 32(1), pp.146-165, January.

Translator's Commentary

In this translation we have cited references to Cassel and Marshall using the published English translations where possible and these references replace those cited by Stackelberg in the German language editions. The pattern of the citation is as follows: "See Cassel (1967): 18". For other citations, the original German, French, etc. edition was quoted, including the page number in that edition and we have indicated the published English title in brackets and italics if this is known e.g. "See also Vilfredo Pareto, (1906) *Manuel d'économie politique*, (published in English as: *Manual of Political Economy*), Paris 1927: 148, No.10". However, in some instances no English translation is known to have been published and so the page references refer to the original German (etc.) edition. Our translation of the unpublished title is given in brackets to help the reader, but is not in italics e.g. see Bücher, "Gesetz der Massenproduktion" (Law of Mass Production) in *Die Entstehung der Volkswirtschaft* (The Rise of the National Economy), 2nd collection, Tübingen 1921: 92, footnote.

There have been very few changes in the English terminology used by our team in this second translation of Heinrich von Stackelberg's work, compared to our first book. The term, "Produktionsmittel" is one of the few exceptions. After consultation, and in an attempt to improve our neutrality in a contentious area, in this second book our translation team has rejected the term "means of production" for its Marxist connotations, although it was used in *Market Structure and Equilibrium* (2011), and has replaced it with "factors of production". On a few occasions we also found the term "capital goods" to be a useful alternative (e.g. Chap. 4, § 2, II, 1.). In Chapter 4, § 3, I, we explain that we have made another change. Stackelberg (1932) uses "Duopol" spelled with a "u" and so that is the spelling we have adopted in English. This is in contrast to the other book (1934) where Stackelberg and therefore we (2011) spelled it "dyopoly" throughout.

Sir Alan Turner Peacock translated Stackelberg's *Grundlagen der Theoretischen Volkswirtschaftslehre* (1948), published in English as *The Theory of the Market Economy* (1952). The translators have purposely avoided reading Peacock's book, but have adopted some of the terminology using the results of a corpus search and these have been clearly cited. However, our terminology differs from Peacock's in one key instance. In Chapter 4, where Peacock used "technical" to translate "technisch" e.g. "technical level" (see Peacock: 6) for "technische Niveau", we have instead preferred "technological" in line with the more recent scholar, Baumgärtner (2001: 513). In contemporary orthodox microeconomics, there is a preference for using "technological" in contexts which concern the production function.

In general terms this translation follows some of the elementary rules for clear writing recommended by the literary author, George Orwell (*Politics and the English Language*, 1946). Firstly, we have avoided introducing metaphors, similes or other figures of speech. Secondly, we have used short rather than long verbs and nouns, apart from where we have used exact terminological equivalents which have a tendency to be longer. We have usually preferred terminology based on Latinates where appropriate, and avoided archaic language with non-native English readers in mind.

We have been disciplined about using consistent terminology and in particular we have benefitted from translation memory software. This consistency of lexical choices throughout the book is intended to help our anticipated readers. Furthermore, we have also streamlined the translation of Stackelberg's rich German vocabulary. Where he uses several synonymous parts of speech (e.g. nouns, verbs, adjectives, adverbs), we have replaced these with one consistent English translation.

In two instances we have rejected Orwell's advice (1946) about how to write in clear and precise English. Firstly, we have continued to use the passive voice in English, echoing Stackelberg, rather than avoiding it as Orwell proposes. This use of the passive in the English translation therefore reflects the publication year of the original book, 1932. Secondly, although Orwell recommended that if it is possible to cut out a word, it should always be cut out, we have avoided this where possible. If we had made subjective choices about which words to exclude, we would have been imposing our views and interfering with the author's intentions. The text would then only have been accessible to the reader through an even thicker translation "veil" than is already an inevitable part of the process. In this sense, the translation is as objective as it can be.

Far from omitting words, on occasion we have inserted words, but we have limited ourselves simply to replacing the author's pronoun that refers to an earlier noun with the relevant noun for clarity and stating the definite or indefinite article: e.g. in Chapter 4, § 4, I where we replace "it" with "this point": "Um ihn zu finden, überdecken wir . . ." becomes, "In order to find this point, we will cover . . ."

It was sometimes also necessary to insert additional words because of the difference between the source and target languages, German and English. Although German adds endings to nouns and pronouns to signal gender and singular or plural so that it is clear what a pronoun is referring to, in English, the equivalent pronoun "it" lacks these markers and can sometimes be ambiguous. Similarly, where Stackelberg was able to elide nouns or verbs, thus exploiting German grammar rules which follow a different word order compared to English, we have added or repeated nouns or verbs in English for improved clarity (see the following underlined repeated verb):

e.g. "Ist .., so hängt die Reaktion des Preises P_1 auf die Veränderung der Menge x_i, also des Angebotes des Gutes Nr. i, von der Nachfrageelastizität dieses Gutes Nr.

i ab" becomes "Where ..., the response of price P_1 depends on the change in quantity x_i, and the supply of good i therefore depends on the elasticity of demand for this good i." One important addition made by the translators was to insert titles for all the Figures to facilitate digital access, apart from the title of Figure 3.1 already included by von Stackelberg. The reader will note that the majority of Figure images are authentic, although it has been necessary to translate the axes labels and add Figure titles to facilitate digital access.

My grateful thanks go to Dr. Bazin as well as Dr. David Urch, Mr. Theodore Goumas and esteemed colleagues of the CIoL and the ITI for their valuable mathematical or linguistic assistance. Our translation team would also like to thank Dr. Gwyneth Mulholland, an associate lecturer in statistics at the University of Ulster, for kindly acting as an additional editor in the process of making the translation as accurate and as readable as possible.

Damien Bazin, Lynn Urch and Rowland Hill

About the Translators

Damien BAZIN, Ph.D. in Economics, is Associate Professor (Accreditation to Supervise Research) in Economics at the University of Nice Sophia Antipolis, France. After completing his postdoctoral studies in Canada and holding a visiting post in China (European Union program), he joined GREDEG (Research Group on Law, Economics and Management) in 2012. He now conducts research in Macroeconomics and Sustainable Development. His fields of specialization are many and varied, including Ecology, Ethics and Environmental Economics, Macroeconomics, Corporate Ethics, and topics related to Socially Sustainable Development. His doctorate research work focuses on responsibility and accountability from an economic perspective.

Damien's Accreditation to Supervise Research highlights a transition from economic ethics to environmental economics. He believes that taxation and sustainability should be viewed through the prism of Corporate Social Responsibility, and that to understand this innovative thinking is to understand the aspects in which a Kantian approach, using motivational variables, might differ from that of business ethics. Following the same line of reasoning, he has demonstrated that both company responsibility within an environmental context and tax-related behaviour must be examined from an ethical standpoint. His interpretation is that taxation is a factor in encouraging responsible behaviour.

His current research deals with the epistemological foundations of ethical economics. He considers the question of environmental ethics and optimal taxation. In doing so, his hope is to create a model for the formal application of responsibility which takes various economic actors into account. As such, he pays special attention to the role of companies and the importance of stakeholders. Given that ethical motivations, as seen from a positive approach, are the mainstay of his analysis, his research has drawn inspiration from social economics. The results of his research reflect the complexities and paradoxes in the premises of social economics. They highlight the probability of undesirable secondary effects arising from crowding-out effects and the emergence of a perverse social capital.

His future papers will cover three areas of research: the ethical stakes in Socially Sustainable Development; behavioural analysis by way of empirical themes (living conditions and modelling of households) using the Capability approach; and the convergence of the previous two areas via well-being, capabilities, the role of the child in development (child labour/street children), agent responsibility, the limits and indicators of Socially Sustainable Development, and lastly, gender inequality and vulnerability. The underlying framework to all this is the construction of policies which optimize the research/expertise interface and which give particular importance to strategies for poverty reduction.

Damien lectures in prestigious Parisian business schools on the subject of company ethics. His most recent papers are related to economic ethics and the protection of nature. He regularly contributes articles to daily national papers with the aim of popularizing fundamental research.

Lynn Urch is a Qualified Member of the Chartered Institute of Linguists, working as a freelance translator of German and French into English. She was educated in translation and modern languages at the University of Hull and the University of the West of England and also has an economics background.

As a young economics student, **Rowland Hill** was inspired by the work of Ernst Schumacher, whose vision of economics was very different from that proposed by the standard textbooks of the time. As a translator, Rowland has maintained his involvement with economics, but in the role of interpreter rather than theoretician. He would like to take this opportunity to dedicate von Stackelberg's book to his former economics teacher from Coleraine Inst in Northern Ireland, Mr Sid Grey, his regular sparring partner in the socialism versus capitalism debate.

Contents

Chapter 1
Foundations

§ 1. Fundamental Concepts of Production

I.

1. We start with the term for the unit of the quantity of goods. As a general rule, this unit can be fixed arbitrarily. However, we want to define it appropriately so that it represents as small a quantity of goods as possible that are typically still sold in the wider market.

We then look at the socio-economic production process. This may be broken down into its individual stages of production so that within each and every stage, the production of a quantity of goods unit is determined by a single purpose. Dividing up this production process is purely theoretical. Normally it will not always be possible to physically divide it up in a clear manner according to this principle. However, this is not necessary for our purposes. No matter how they are physically divided up, the propositions we will deduce in this section remain valid because they are also formal in nature.

Services are dealt with in exactly the same way as physical goods. We can also allocate units to them and we regard the provision of these units as being production. Hence, if services are to be considered as being the aim of production, they do not need to be mentioned as a special case hereafter.

The firm's production is all of the resources used for the productive purpose of achieving production within that stage of production. We describe the finished goods produced by a firm as its product.

2. Furthermore, we divide up the production process into sections in which production is dependent on a single economic interest. We refer to these sections as individual branches of economic activity. They may simply comprise entire stages of production, or even one stage or several. The firms belonging to one branch of economic activity constitute an enterprise so long as they have the same economic interests.

H. von Stackelberg, *Foundations of a Pure Cost Theory*,
DOI 10.1007/978-3-642-34537-1_1,

Now if we look at the whole production process for an entire category of goods, we will see that several firms at the same level can exist within each stage of production. An enterprise can therefore also be comprised of multiple firms within that same stage of production. Similarly, multiple enterprises can also exist in the same branch of economic activity. Moreover, dividing them up into stages of production and branches of economic activity does not necessarily need to be standardised for the same category of goods because two firms that are combined vertically can be regarded as one firm.

3. The goods and services which a firm requires to manufacture its products are its factors of production (Translator's note: "factors of production" see Cassel (1967): 18).

A. Producer goods

Producer goods, like all goods for that matter, can be divided into consumption goods and durable goods.[1] We describe the former as operating supplies and the latter as operating equipment.

1. Operating supplies enter the product as a quantity by way of production. Therefore, these are raw materials and supplies as well as similar factors of production, such as power, if these are purchased from other firms.
2. We cannot talk about operating equipment entering the product. Instead, it is a more or less durable production process. It is the equipment's services, rather than the equipment itself which enter the production process. Its service life,[2] that is, the period of its useful life, can depend on either the time period or on the extent to which it is used for production, or alternatively on both or on neither of these factors.

Accordingly, we can distinguish between the following kinds of operating equipment:

(a) Land: by this we mean the forces of nature and the land's natural resources that are linked to a particular location (in an abstract sense, we cannot add to or impair these natural resources, although they may be self-renewing) and the production advantages linked to this location that result from the development of society as a whole. The service life of the "land" does not depend on the time period or degree of use.
(b) All other functional factors of production, so long as they are not operating supplies, that is, e.g. buildings, machines, tools, apparatus, etc. These depend on the time period as well the degree of use for their service life.
(c) Certain rights, such as e.g. copyright, patent rights and similar. These only depend on the time period for their service life.

[1] See Cassel, G. (1967): 11 et seq.

[2] See Schmalenbach, E. 104 and 113.

(d) Stocks of consumable goods, that is, primarily natural resources and similar. The service life of consumables depends solely on their use, that is, on the materials that are consumed. On the face of it, these capital goods become operating supplies. However, they must be included in operating equipment calculations because life span is crucial for this kind of equipment.
(e) Fixed stocks, termed "base stock". These must also be counted as operating equipment because of their life span. However, their service life span is arbitrary. Therefore, they ought to be termed "intangible operating equipment". They have a certain similarity with the "permanent participation" of human labour in the firm which is discussed below.

B. Productive services

We divide these into operating equipment performance and the work performed.

1. Operating equipment performance has to be a component of the firm we are examining, and can therefore be classified accordingly.
2. We divide the work performed into two categories:
   (a) The first category is work purchased individually as such by the firm so that the workers themselves are not regarded as being directly employed by the firm. This work is similar to operating supplies in its relationship to the firm and mostly corresponds to the work performed.
   (b) The second category comprises work undertaken by people who can be regarded as being employed by the firm. This work now seems to be part of a more or less constant "participation" in the firm on an employed basis. This permanent participation is similar in nature to operating equipment, especially to "fixed stocks" (see e), Section A). Therefore, it mostly concerns managerial employment. However, manual work must also occasionally be included e.g. where skilled or semi-skilled workers are trained for complex production.

We now want to outline an important division of the factors of production. We have seen that one part of the factors of production immediately goes into the production process. We can in fact measure the participation of this factor of production in the production process over a specific time period. We can establish how many operating supplies are going into the production process, the level of operating equipment performance throughout this time and how many man-hours have been worked. We can describe the factors of production characterised here as direct factors. We cannot estimate an additional aspect relating to the factors of production in this way in terms of its importance for production. We also cannot determine the importance for production of the operating equipment (irrespective of its performance) throughout the time in question nor are we are able to do this for "permanent participation" (irrespective of its individual performance). The aim of this remark is not to split hairs. It is in fact possible, for example, to obtain the same (technical) performance over the same time frame from two very different machines which are also very different in terms of size. We are therefore entitled to ask about the difference between the uses of these two machines, irrespective of their actual performance at any given moment.

We combine operating equipment and "permanent participation" using the term "indirect factors of production". However, here we want to isolate "permanent participation" and "fixed stocks" and refer to them as "intangible indirect factors of production" because, as was mentioned earlier, they are different in some respects from the other indirect factors of production.

We have therefore established the following division:

I. Direct factors of production:

   1. Operating supplies;
   2. Operating equipment performance;
   3. Work performed.

II. Indirect factors of production:

   Operating equipment (except "fixed stocks").

III. Intangible indirect factors of production ("fixed stocks" and "permanent participation").

This difference between the direct and indirect factors of production has an important role to play in cost theory.

II.

1. "An economic consideration of production must start from the fundamental fact . . .that. . . production is a continuing process.[3]"

This conclusion should also be used as a basis for examining the production carried out by an individual firm. An arbitrary production quantity is produced by the firm within a period of time. Production is the provision of a particular quantity within a particular time. Hence production has a particular velocity. We measure the velocity of production (Translator's note: "velocity of production" see Baumgärtner, Stephan, (2001), "Heinrich von Stackelberg on Joint Production", *The European Journal of the History of Economic Thought*, 8(4): 514) of a firm using the quantity of a particular type of product currently produced in the unit time by the firm.[4]

A firm can usually achieve different velocities of production. The velocity of production is therefore regarded as being a changeable variable. Furthermore, where multiple categories of products are produced, one velocity of production is achieved

---

[3] Cassel, G. (1967): 25–26.

[4] See also Pareto, V. (1906) No. 10.

for each category of product. The "production level" of a firm is only determined in this case if details of the velocity of production are supplied for each individual category of product. For example, if three different types of goods are produced, then to determine the firm's "production level" we must be given details of the velocity of production for each individual type of good, that is, three numbers must be given overall. To sum up, we denote these numbers via which the production level of a firm is determined in the case of joint production by a product vector (and this is also in line with mathematical terminology). In addition this product vector is usually variable because each and every one of the velocities of production contained in it is variable, or, as we could also say, each individual component is usually variable, that is, a given firm can usually achieve very different production levels (technologically) and can produce each of its product types at different velocities. This then results in the firm combining its factors of production (qualitatively and quantitatively) in various ways.

It should be noted that the factors of production are also used on a continual basis within the period of time. We can regard this use to be a special kind of consumption. In fact, it is the consumption of economic goods. This use is referred to as the input and because this occurs as a continual process throughout the period of time, we can now also talk about the velocity of input. And indeed, a velocity of input is assigned to each of the individual factors of production (as a changeable variable). Now we can talk about an "input level" which is characterised by specifying all of the velocities of input. We combine the individual velocities of input for an input level into an "input vector".

Production is presented as obtaining an input vector for the purposes of obtaining a product vector. However, the firm's economic situation is still not determined by simply giving an input vector. We need the time allocation for the expenditure on the individual types of the factors of production, or, as we can more easily term this, the period of production (Translator's note: "period of production" see Alan T. Peacock, (1952), *The Theory of the Market Economy*, London, Edinburgh, Glasgow, William Hodge Company limited, XXIII+318 p. (transl. *Grundlagen der Theoretischen Volkswirtschaftslehre* by H. (von) Stackelberg): 75). Each individual type of factor of production is only expended over a unit time in order to obtain a particular velocity of production (to observe the simple case of producing only one good) and hence to produce a particular quantity of products in the unit time.[5] However, the expenditure on the various factors of production can be distributed over many different periods of time (Translator's note: "periods of time" see Alfred Marshall, (1961), *Principles of Economics*, 9th (variorum) edition, with annotations by C.W. Guillebaud, volume II, Notes, Cambridge, MacMillan and Co. limited for the Royal Economic Society, 886 pages: 342), either spread over long periods of time or condensed over short periods. We can therefore obtain a velocity of production using different periods. An input vector that leads to the achievement of a particular production level within a particular production period is not appropriate for a different (perhaps shorter) production period in other circumstances.

---

[5] If we in fact assume that production occurs continuously.

The production period plays an important part in determining a firm's capital requirement.

2. In the first part of this section we examined the fundamental division of direct and indirect factors of production. We must now focus our attention on the significance of this distinction by analysing production.

We must be clear about what we mean by the input of a direct factor of production and the input of an indirect factor of production.

The input of a direct factor of production is indeed a very important concept, given our remarks about the comparison between direct and indirect factors of production. It is the quantity of the factor of production units for the type of factor of production concerned which has entered the production process during the period of time concerned. The definition of the velocity of input of the direct factor of production is similarly also very clear.

By contrast, the term "input" is more difficult to define for indirect factors of production. There is no "input into the production process" in this case. Instead, indirect factors of production form a permanent basis of production. We can determine the scope of this basis, and hence the quantity of the indirect factors of production associated with the firm. Furthermore, we can also calculate either the increase or decrease in the indirect factors of production to or from the firm within a given period of time. The decrease results from consumption. If the production basis which is described using the indirect factors of production does not change, then the relevant increase occurring during a period of time must be equal to the consumption of the indirect factors of production occurring during this period of time. Consequently, we would have a model for the consumption of the indirect factors of production in the case of an increase which is always necessary for maintaining an unchanged basis of production. However, this consumption not only results from the input of the indirect factors of production themselves, that is, from the input of the factors of production that have been brought into the firm, but also from the input of the services of this indirect factor of production, that is, from the cost of the direct factors of production. Consumption must be divided and allocated. It is usually divided by allocating consumption to the indirect factors of production which would exist if the firm were to lie idle, that is, if the expenditure on the direct factors of production were to stop. The difference between this consumption and the consumption for any production level is then allocated to the performance of the relevant indirect factors of production.

We can therefore distinguish:

1. The input of the indirect factors of production, which means bringing indirect factors of production into the firm as a permanent basis of production. We shall refer to this input as an "indirect input" in its simplest terms.
2. The input of the direct factors of production. This involves indirect factors of production on a given basis. We shall refer to this input as the "direct input" in its simplest terms.

The firm can adjust direct input much more quickly and easily than indirect input. The firm changes its production level in the first instance by changing direct input, and then by changing its indirect input in the second instance. The combination of indirect factors of production is always determined for a longer period than the combination of direct factors. The enterprise also responds to short-run economic changes simply by altering its direct input. This input only leads to economic change in the long run and it also alters indirect input. These facts will become even clearer later.[6]

From these observations we can see that the problem of the change in input consists of two issues. Firstly, we can study the change in direct input, that is, the various activity rates of the firm with a constant firm size (expressed rather imprecisely) and secondly, we can either study changes in indirect input or changes in the size of the firm itself.

III.

1. So far we have just dealt with the issue of quantities. We must now deal with the value of these quantities and indeed with their objective exchange value, that is, their money valuation ("money valuation" see Cassel Gustav, (1967), *The Theory of Social Economy*, translated by S.L. Barron, New revised edition, New York, Augustus M. Kelley Publishers, coll. Reprints of Economics Classics, VIII+689 pages: 49). For this purpose we introduce an ideal money scale (Translator's note: "money scale" see Cassel (1967): 50) in the sense described by Cassel in what we regard as a social economy.[7]

Next we establish that production is linked to a consumption value. The quantity of factors of production consumed in a period of time in order to realise a particular price level has a money value which is the result of multiplying these quantities by the corresponding prices to obtain the total.

Lastly, the consumption value results from the necessity of keeping the indirect factors of production constant. To be able to do this, the enterprise concerned must permanently have a specific purchasing power at its disposal. There is a cost attached to this availability of capital and consequently a consumption of assets which is equal to the price of this availability of capital and hence to the interest rate in the unit time. The capital required is equal to the value of all the indirect factors of production of the enterprise. We refer to this as either standing or fixed capital.

Ultimately, the enterprise needs a further availability of capital to eliminate the production period already mentioned above. By expending a specific quantity of the factors of production, capital is determined in the unit time. This capital remains fixed until the product for which the expenditure concerned was incurred to produce this product is sold. The value of the quantity of expended factors of production

---

[6] In addition, see Alfred Marshall, Theorie der Quasirente (Theory of Quasi-rents): Marshall (1890).

[7] Cassel, G. (1967): 44.

multiplied by the period of time of the expenditure up to the point where the product is sold, results in the capital requirement caused by expending the quantity of factors of production concerned. From this, it follows that we obtain the capital requirement for an input level (for the given production period, or to be more precise, for the given allocation of the expenditure of the individual types of factors of production in the period of time) by adding up the resulting capital requirement variables for each individual type of factor of production for the relevant input level. We hence obtain the amount of so-called circulating capital or working capital. This also results in a consumption value that is equal to the interest rate in the unit time.

By obtaining an input level for a particular production period, we define the total existing consumption value to be the total costs of this input level.[8]

An input level for a particular production period is obtained in order to achieve a particular production level. There is one input level which features the lowest total costs of all the cost levels suited to achieving a particular production level. We describe the total costs of this input level as being the total costs of the production level concerned.[9]

These total costs mean something different depending on whether they are calculated generally according to all of a market economy's possible input levels or whether they apply to an actual firm. In fact, in the latter case, not all of the possible alternative cost levels are achievable. The indirect input is assumed to be invariable if the time period in which the relevant production level will be achieved is short. The longer this time period is, the better the firm can also adjust its indirect factors of production and the more likely it will be that the input levels in the relevant time period will also match all potential input levels. When total costs are mentioned in the following pages, it will always be with reference to a firm's total costs and to the assumption that the indirect input is invariable, providing nothing else is expressly stated.

2. Obtaining a particular production level, or if we express it another way, obtaining a particular product vector in an enterprise, requires a particular level of total costs. Let us examine a simple case where only one good is produced. We can then also say that a particular level of total costs is required to obtain a particular velocity of production, that is, a particular level of total costs is associated with a given velocity of production and these total costs must arise in the unit time and must be borne so that this velocity of production can be achieved. In other words, total costs (which must always be applied to the unit time) are either a function of the velocity of production, or in the more general case, are a function of the product vector.[10] This function is unambiguous. No velocity of production can possess multiple levels of total costs which might be different from each other

[8] See Cassel, G. (4th Ger. ed.): 77. See also Amoroso, L. 1930: 2, definition of "Costo totale" (Total Cost).

[9] In this way, every input vector corresponds to a product vector. The partial derivatives of the velocities of input, according to the velocities of production, are "technical coefficients" as Pareto termed them (coéfficients de production). See Pareto, V.: 607, equations (101). Also see Pareto's explanations about the variability of technical coefficients, ibid.: 326 et seq., No. 70.

[10] Here, we will use the Dirichlet function.

because there is a lowest total cost among all of these total costs. The other total costs are then ruled out according to the definition of the term "total costs".

We initially limit ourselves to a case in which only one good is produced. Now we can determine another feature of the total cost function: it is a monotonically increasing function with an increasing velocity of production. A higher velocity of production cannot then have lower total costs than a lower velocity. Since the lower velocity is included in the higher velocity, it can hence simply be obtained by achieving the higher velocity. This assertion can also be extended to a generalisation about joint production. A product vector cannot have higher total costs than another product vector if none of the first product vector's components is greater than the other's corresponding components.

In this way, we have derived two basic features of the total cost function, that is, it is unambiguous and it increases monotonically. The main purpose of the following investigation is to establish other features of the total cost function and draw conclusions from these features.

3. Since we sometimes need to express mathematical concepts, we will introduce mathematical symbols for some variables at this point. Therefore, from the outset we will distinguish between single production (Translator's note: "single production" see Baumgärtner: 122) and joint production.

(a) Single production.

Let x be the velocity of production of the good produced.

Let K be total costs. We also use this symbol in the text because the term "total costs" is a plural form and is therefore awkward. Total costs appear to be a function of x. To show that a variable is a function of another variable, we put the second variable in brackets after the first, as is usual. Consequently, we can write:

$$K = K(x).$$

Let P be the price of the good produced, that is, the quantity of money paid for the unit of quantity of the good in the market.

(b) Joint production.

Here, multiple goods are produced. We number the goods from 1 to n. The velocities of production are assigned the number of the corresponding product as an index. The velocity of production of good 1 is x_1, good 2 is x_2 etc. A product vector is a system of n velocities of production:

$$(x_1, x_2, x_3, \ldots . x_n).$$

We refer to this vector using $\mathfrak{x}$.

Total costs keep the symbol K but they are now dependent on n velocities of production. We then derive the following:

$$K = K(x_1, x_2, \ldots .. x_n).$$

Since an invertible and definite covariance matrix exists between a system of n velocities of production and the corresponding product vector, we can hence also state that:

$$K = K(\mathfrak{x}).$$

Every good produced has a price. We use the number of goods in the same way we did for velocities of production by assigning a number as index to the price of a good. Therefore, good 1 then has price P_1, etc. We thus obtain a system of n prices which we term the "price vector". We introduce the letter $\mathfrak{P}$ to be the symbol for the price vector.

We will introduce other symbols when required in the course of this description.[11]

4. Total costs are the result of the expense associated with direct and indirect inputs. The indirect input is the same for each production level. This also applies to fixed capital. This is dependent on the indirect factors of production required. As this remains unchanged, fixed capital is hence also unchanged. The difference between the total costs for two different production levels is determined by the difference between the expense associated with the respective direct input and the working capital required.

As a result, we can imagine total costs as being composed of two components added together, one constant and one variable. We describe the first component as "constant costs" and introduce these costs using the symbol K_{I}. We describe the second component as "variable costs" and denote these using the symbol K_{II}. Only K_{II} is dependent on the production level, whereas K_{I}, as mentioned earlier, remains the same for all production levels.[12]

However, we must now consider that the expense associated with indirect input should also be regarded as partly variable for longer periods of time (as indicated above). As a result of this, the picture would change if a greater share were to be allotted to variable costs. However, for short periods, the expense associated with indirect input, as outlined a moment ago, applies, as well as that between variable costs and the expense associated with direct input.[13]

Furthermore, we can distinguish another cost type. These are costs that change by jumps and then remain constant for all those production levels which remain continuously interrelated. The simplest way to look at this situation is in the case

---

[11] The condition of monotonicity in the narrowest sense will be formulated mathematically as follows: in the case of single production the following always applies: $K(x) < K(x')$, if $x < x'$. In the example of the joint production of, for example, two goods, the following always applies: $K(x_1, x_2) < K(x'_1, x_2)$ and $K(x_1, x_2) < < K(x_1, x'_2)$ if $x_1 < x'_1$ and $x_2 < x'_2$.

[12] We can then state the following:

$$K(x) = K_{\mathrm{I}} + K_{\mathrm{II}}(x), \text{ or}$$
$$K(\mathfrak{x}) = K_{\mathrm{I}} + K_{\mathrm{II}}(\mathfrak{x}).$$

[13] Constant costs are primarily Schmalenbach's "fixed" costs whereas variable costs correspond to Schmalenbach's other cost categories. See Schmalenbach, E. (1930): 32 et seq.

where only one good is produced. It is these costs that maintain a constant level up to a specific velocity of production; they then jump but are then constant again for the next interval before perhaps jumping again, etc. We shall describe these as jumping costs (Translator's note: "jumping costs" see Heinrich von Stackelberg, Political Events and Economic Ideas: 255) or semi-variable costs and introduce the symbol K_{III} to denote them. They are dependent on the activity rate so we must write them as $K_{\mathrm{III}}(x)$. They are hence variable costs in principle. However, it can sometimes be useful to separate these jumping costs and to combine all the variable cost points of discontinuity using this new function, that is, to summarise total costs overall so that variable costs in the narrowest sense, and as a result, total costs minus the jumping costs too, would then be continuous functions of the velocity of production. Due to monotonicity, total costs and hence also variable costs (which are similarly monotonical) can only possess points of discontinuity if they jump upwards. However, this circumstance becomes complicated for joint production. Therefore we do not wish to continue to pursue it now.

The source of jumping costs lies primarily with intangible indirect factors of production,[14] but it can also result from direct input.

Hence, total costs are composed of three summands.[15]

We still have to determine how constant costs can be calculated if the total cost function is known. This is very simple. By definition, constant costs are invariable for all production levels. Now however, there is one production level at which variable costs (including jumping costs) have a zero value. This is the production level at which absolutely nothing is produced, i.e., if the firm lies idle. As the expense associated with direct input is in fact arbitrarily variable, it is not incurred at all if the firm lies idle. This is because the lowest costs arise for this production level. The same also applies in certain circumstances for part of the intangible indirect factors of production. However, from this it emerges that constant costs are equal to total costs if all velocities of production have a zero value, that is, if $x = 0$ or $\mathfrak{x}$ is a zero vector.[16]

Unless there are significant changes, we can also define K_{I} in a different way by equating K_{I} with the lower limit of all values of $K(x)$ or $K(\mathfrak{r})$, for which it follows that $x \neq 0$ or $\mathfrak{r} \neq \{0\}$. The meaning of this is especially clear for the case where only one good is produced. Here, this lower limit is just the marginal value of $K(x)$ if x converges to 0.[17] Where $K(x)$ is discontinuous at the origin, K_{I} now has a different

[14] Now, let us remind ourselves of a famous example in management studies: a bookkeeper can perhaps post up to 1,000 book entries a day. If there are 1,010 entries to post in the firm, a second bookkeeper must consequently be employed and so there is a sharp increase in total costs. The bookkeeper's work relies on "permanent participation".

[15] $K = K_{\mathrm{I}} + K_{\mathrm{II}}(x) + K_{\mathrm{III}}(x)$.

[16] The following equation hence applies:

$$K_I = K(0), \text{ or.}$$
$$K_1 = K(0, 0, \ldots, 0) = K(\{0\})$$

if we denote the zero vector using $\{0\}$.

[17] Then we would have: $K_I = \lim\limits_{x \to 0} K(x)$.

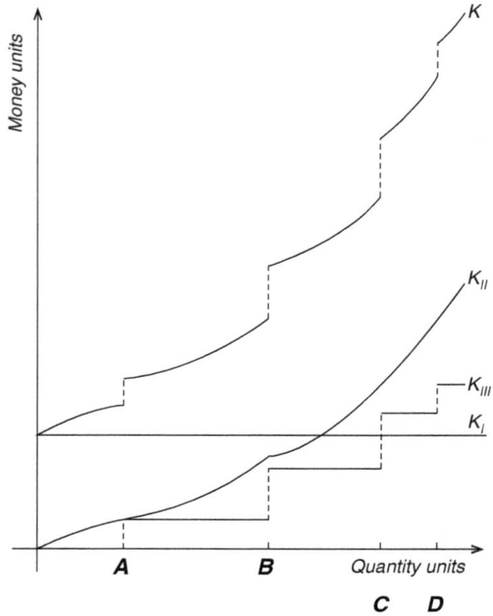

Fig. 1.1 A general total cost function

value compared to that in the first definition. We can introduce terms for $K(0)$ and $\lim_{x \to 0} K(x)$ that are linked to a certain notion. $K(0)$ are idle time costs and $\lim_{x \to 0} K(x)$ are production readiness costs.[18]

This different definition is not of principal significance. Even so, we would like to make constant costs equal to idle time costs in the following pages for reasons of efficiency.[19] This is because the formulation of the costs we derive consequently becomes more standardised. In order to prevent misunderstandings, it should be noted that in our context, we are primarily referring to short run idle time costs. We have seen that the difference between cost types is dependent on the period of time the enterprise carries out production regulation. Similarly, idle time costs also differ according to the period of time concerned. Since we are explicitly concentrating on very short periods of time, the term "idle time costs" is hence also to be understood accordingly.

The difference between "idle time costs" and "production readiness costs" will be of no importance for the fundamental parts of our explanations because we assume that the functions considered are continuous and can be differentiated. This is important when we generalise about costs. Only jumping costs also occur here.

Fig. 1.1 is a graph that illustrates, for example, how a general type of total cost function should appear and what it should consist of.[20]

---

[18] According to Schmalenbach, E. (1927): 31.

[19] Hence $K_1 = K(0)$.

[20] See also: Peiser, H. (1924): 10.

Curve K_I represents constant costs. Accordingly, it is parallel to the quantity axis. Curve K_{II} represents continuously variable costs. In accordance with its definition, it is continuous, that is, it never jumps. Curve K_{III} represents jumping costs. It only changes where total cost curve K jumps. Total cost curve K results from the sum of the ordinates of the three marginal cost curves. At the four points A, B, C, D, it is a monotonically increasing function, and in our example, a discontinuous function. Continuously variable cost curve K_{II} is parallel to total cost curve K but is shifted downwards by constant and jumping costs.

§ 2. The Societal Determinants of Production

I.

We, along with Cassel, must acknowledge the principle of scarcity (Translator's note: "principle of scarcity" see Cassel: 168) to be the first and most important determinant of the economy and hence also of production as a partial function of the economy. Cassel explains this principle in detail in the first paragraph of his book, so we only need to cite him here. We will content ourselves with citing the definition Cassel gives in another part of his book for the principle of scarcity, *The Theory of Social Economy*. He says[21]

> "... it is the principle of scarcity that is lost sight of - that is to say, the significance of the limitation of the available factors of production".

II.

The second principle which we now examine is the "economic principle". Cassel also explained this in the first few paragraphs of *The Theory of Social Economy*. Only one aspect of this principle, referred to by Cassel as the "principle of scarcity" (Translator's note: "principle of scarcity" see Cassel (1967): 5), is possible for production. However, we must stress here that this principle is significantly different from the technical principle of minimum means (Translator's note: "principle of minimum means" see Cassel (1967): 7). It has stricter selection criteria in terms of the opportunities for achieving a purpose compared to the technical principle of minimum means. This latter principle must ensure that the aim is achieved, but by

[21] Cassel, G. (1967): 170 (Translator's note: this is the exact quotation taken from Barron's English translation).

expending the smallest possible amount on the means. This partial principle taken from the economic principle mentioned above, which we can also describe as the "economic principle of minimum means", must by contrast ensure that the aim of expending the lowest money value possible is achieved. Since a larger quantity cannot have a lower value than a smaller quantity, this technical principle of the minimum means is hence contained in the economic principle of minimum means. It also follows that if only one factor of production is used, then the economic principle agrees with the technical principle, but if multiple factors are used, which may be assumed to be the general rule, this economic principle then goes further. The technical principle actually only selects those combinations of resources which would also be selected using the economic principle. However, there are still combinations that are indifferent in terms of this technical principle, whereas other combinations may be selected using the economic principle because the individual factors are reduced to their common denominator, even though they cannot be physically compared or added together.

We can examine this in greater detail by representing individual combinations of factors as input vectors in concrete terms. According to the technical principle, two equally strong[22] input vectors are indifferent. However, according to the economic principle, one of these two vectors is ruled out if its costs are higher than the other's.

In our opinion this difference between the technical and the economic principle has not been clearly demonstrated to a sufficient degree by Cassel. On the basis of our earlier explanations, we can formulate the economic principle as it appears in this study as follows: the economic principle demands that a given purpose is achieved with the cheapest means possible, that is, with the least possible input in money terms. On the basis of this definition we can see that our term for "costs" agrees with the economic principle. Anyone who orientates themselves towards this principle will always aim to carry out production at the lowest possible input level. Its reduction in value will hence tend towards our "production level total costs" and may consequently be treated as total costs. It even exhibits the same features.

III.

The production aim is set by the director of the enterprise or the entrepreneur.[23] There may be a wide range of objectives, but we shall limit ourselves to our own experience here. We take two clearly distinguishable objectives which we

---

[22] We can say: "a vector a is weaker than another vector b" if none of the components of vector a are greater than the corresponding components of vector b. We can then write: $a << b$. We can then also say: "b is stronger than a" and write that $b >> a$. If neither $a << b$ nor $b << a$ applies, we can hence say that "the two vectors a and b are equally strong" and write: $a \sim b$. If $a << b$ and $b << a$ apply simultaneously, then the two vectors are equal to each other: $a = b$.

[23] We are using the word "entrepreneur" in a technical sense, not for instance in the sense of an entrepreneur in a capitalist economy.

contrastingly term the "profit-making principle" (Translator's note: "profit-making" see Peacock: 291) and the "principle of the satisfaction of needs and wants" (Translator's note: "satisfaction of needs and wants" see Michael Wohlgemuth, (2004): The Communicative Character of Capitalistic Competition: A Hayekian Response to the Habermasian Challenge, Freiburg Discussion Papers on Constitutional Economics, No. 04/1, http://hdl.handel.net/10419/4334: 12).

1. The entrepreneur behaves according to the profit-making principle if he sets profit maximisation to be his production aim. We must comment on this proposition in greater detail. The product produced in the unit time is sold in the market. Its price multiplied by its quantity is the revenue of this good in the unit time, or, as we would prefer to term this from now on, its return. The sum of the returns of all goods produced by the enterprise is the return of the enterprise (in the unit time). We describe the difference between return and costs for a given period of time as the enterprise's profit or loss within this period of time. It can be positive or negative.

Striving for the greatest possible profit therefore means striving to achieve the greatest possible difference between return and costs during a particular time. Using simplified assumptions, profit is maximised during this time if the profit in the unit time is maximised throughout this entire time. The purpose of striving in the above manner is hence to achieve the greatest possible profit in the unit time. It is in this sense that we would like profit to be understood from now on.

In order to achieve his aim, the entrepreneur must behave in accordance with the economic principle.

2. The entrepreneur behaves according to the principle of the satisfaction of needs and wants when he sets his production aim to meet the supply of any ordered production quantity as cheaply as possible. This proposition also requires an explanation. The word "met" (Translator's note: "gedeckt" in German) is meant to indicate that all costs arising from production (including, for example, from the entrepreneur's wages) must be met by the return. Here, this is treated as "livelihood" rather than profit.

This aim also demands that the economic principle is respected.

3. We find both of these principles in practice. In the pre-capitalist economy, the second might have been dominant. Today, the first dominates. However, the second, as we will see later, has not just become meaningless. Instead, it appears as if the principle of the satisfaction of needs and wants is gaining ground again.[24]

In the following pages we shall focus on these two principles and refrain from discussing any other objectives further.

IV.

The enterprise's market position appears to be the last societal determinant of an enterprise's production which we need examine. In formal terms, we can define this

[24] See e.g. Schmalenbach, E. (1928).

market position as being the relationship between the achievable price and the saleable quantity which the enterprise faces in its market. We distinguish three possibilities.

1. The enterprise may supply in a market where a large number of other enterprises that are independent of one another supply the same good. We are talking here about free competition. If we assume the number of these enterprises is very large so that an individual enterprise's production is close to negligible, we can then say that this enterprise's achievable supply price is independent and that at this price, it can sell any arbitrary quantity of products it is able to produce. Conversely, the above features appear to be a definition of the competitive economy (Translator's note: "competitive economy" see Peacock: 298). This means that enterprises which are to function in this competitive economy could really also exhibit any features which are irreconcilable with these conditions. Later, it will become clear exactly what this means (Chap. 2, § 4).

2. The other possibility is that the enterprise faces a market where its supply plays a significant role because it produces all the supply which reaches the market, or the largest share of it. Here, price is dependent on the quantity supplied by this enterprise. And indeed the established price law applies, namely that price decreases when supply increases and vice versa. This then means that the price which can be obtained is a monotonically decreasing function of the quantity supplied. Since the quantity supplied is identical to the quantity produced and hence in the unit time it is identical to the velocity of production x, we can state the following:

$$P = P\,(x).$$

A generalisation arises in the case of joint production if it is assumed that the price of a good depends not only on the supply of this good but also on the supply of other goods. On the basis of the Dirichlet function definition, we can now formally regard the price of a good as a function of the quantities supplied of all the goods produced by the enterprise, therefore, a function of the product vector. For the price of good 1, we then derive:

$$P_1 = P_1\,(x_1, x_2, \ldots, x_n) = P_1\,(\mathfrak{x})$$

and so on. In total, this makes n functions of n variables.

Here too, the price of a good, for example, price P_1 of good 1, is a monotonically decreasing function of quantity x_1 of this good. Its reaction to other (2–n) goods produced by the enterprise, is different.[25]

---

[25] To summarise, we can for example make the following distinction:

1. Where good i (where i can be $= 2, 3, \ldots. n$) is a rival method for satisfying wants (Translator's note: "satisfying wants" see Cassel (1967): 5) compared with good 1, price P_1 is then a monotonically decreasing function of x_i because an increasing supply of good i decreases price P_i of this good. As a result, demand for good 1 is lured away to rival good i which results in a decrease in price P_1 of good 1 with an unchanged supply of good 1. The same applies to a reduced supply of good i.

The price of any good, that is, a particular price vector, always corresponds to a product vector. In analogy to the relationships between simple number variables (between scalar variables, as they are termed in vector calculus) we can therefore show the price vector as a function of the product vector. We then derive:

$$\mathfrak{P} = \mathfrak{P}(\mathfrak{x}).$$

This equation replaces the n equations outlined above.

3. A third possibility, which often closely resembles reality, still remains to be examined. In a certain sense, it is related to the first possibility, competition, but it exhibits important distinctive features. Here too we find supply divided between multiple enterprises that are independent of one another. A single price corresponding to aggregate demand exists in the market. However the sales each enterprise can obtain are not random in practice, as they were in the first example. Instead, aggregate demand is distributed between the individual enterprises in a more or less fixed proportion. This proportion arises from various social factors, which are often "alogical" even in Pareto's[26] terms. Habit, reputation, advertising, personal relationships and ultimately chance in general are key here. The proportion is hence independent of the price of each individual enterprise's quantity supplied as in the first example. However the saleable quantity is not arbitrary but relatively fixed. And indeed we shall assume that the enterprise is in a position to increase its sales by expenditure on costs e.g. using its sales force, advertising etc. to increase its share of aggregate supply. Similarly, let us assume that sales decrease if the enterprise does not maintain its sales promotion. The sales volume is now dependent on one particular share of costs, sales promotion costs alone, and indeed it is a monotonically increasing function of these costs. We can examine promotion costs and divide them

2. Where good i (i can again be $= 2, 3, \ldots n$) is a complementary good compared with good 1, price P_1 is hence a monotonically increasing function of x_i because an increased supply of good i causes a decrease in the price of this good. As a result, the opportunity to satisfy needs which is jointly provided by the two goods 1 and i becomes cheaper and this leads to a demand for both goods. If quantity x_1 of good 1 remains the same, this can only be similarly restricted if the price of good 1 increases. The same applies if the supply of good i decreases.
3. Where good i is neither a rival nor a complementary good for good 1, the response of price P_1 depends on the change in quantity x_i, and the supply of good i and therefore on the elasticity of demand for good i. Where this elasticity is greater than 1, a larger share of total revenue for good i is hence used than before for an increase in the supply of good i. As a result, only a smaller share of total revenue is left for the other goods, and hence for good 1 too. Therefore in the new situation, the same quantity of good 1 can only be sold at a lower price. P_1 is hence a monotonically decreasing function of quantity x_i of good i in this case. Where the elasticity of demand of this good is lower than 1, P_1 is a monotonically increasing function of x_i which is clear on the basis of similar observations. Consequently, where the elasticity of demand for good i is exactly the same as for good 1, the price for good 1 is independent of the supply of good i.

[26] Pareto, V. ibid., chap. II, 1 and 2 ibid.

up in exactly the same way as production costs. We wish to describe them using the symbol C to distinguish them from tangible production costs K. We then derive the following functional relationships for the case of single supply:

$$P = \text{constans};\ x = x(C);\ K = K(x).$$

Since x is a monotonically increasing function of C, we can rearrange this function and write $C = C(x)$. C is hence a monotonically increasing function of x. All the costs that are tied to velocity of production x are costs of production and sales (Translator's note: "costs of production and sales" see Baumgärtner: 514), that is, $K + C$. They are a function of velocity of production x which shows exactly the same features as function $K(x)$ showed in our earlier observations. Moreover, as price is constant, a case now essentially exists which is not fundamentally different from the general case of free competition as considered in section I. However, we have ruled it out because the given situation here is not readily transparent but can be traced back to the general case using the transformation which we described earlier. In fact, the costs of a good are often discussed in terms of the production costs in the narrowest sense rather than including the sales costs too. From an economic point of view, we must regard sales costs to be part of production costs. Now it is not quite simply a good that is produced, but a marketable good. This fact is important. It may for example often be the case that the production costs follow the law of increasing returns[27] in the narrowest sense, while sales costs are subject to the law of diminishing returns to such a degree that the total costs of the good concerned are similarly governed by the law of diminishing returns,[28] but not as strongly as sales costs.

The situation is similar in the case of joint supply. Here, the sales of a particular product vector $\mathfrak{x}$ can only be achieved with the expenditure on particular sales costs C. C is hence a monotonically increasing function of $\mathfrak{x}$ and indeed for the same reasons as K. Therefore, we now derive the term $K + C$ for aggregate costs. Price vector $\mathfrak{P}$ is constant. It is not possible here to present the product vector as a function of C, similar to that given in the case of single supply, even as a starting point for the presentation of x as a function of C, because the monotonicity is not sufficient to definitely determine $\mathfrak{x}$ for a given C.[29] We can use the term "modified competition" to describe the market situation here.

---

[27] See Chap 2, § 1 below.

[28] See Harrod, R.F. (1931): 566 et seq., as well as the extracts cited there. See also: Allen, R.G.D. (1932): 323 et seq.

[29] However, with the expenditure of a particular level of sales costs, the saleable product vectors of n dimensions represent a great diversity of the range $(n-1)$, given by the equation $C_0 – C\ \mathfrak{x} = 0$, where C_0 is predetermined. Generally speaking, in the case of single supply between C and x and in the case of joint supply between C and $\mathfrak{x}$, a relationship in the form of $\varphi(x, C) = 0$ or $\varphi(\mathfrak{x}, C) = 0$ can easily be established where the derivative $\frac{\partial \varphi}{\partial c}$ is other than zero so that C can always be represented as an explicit function of x or $\mathfrak{x}$.

References

Allen, R.G.D. (1932). Decreasing costs: a mathematical note. *Economic Journal*. 323 et seq.

Amoroso, L. (1930). *La curva statica di offerta* (published in English as: *The Static Supply Curve*). Giornale degli economisti, 2.

Cassel, G. (1967). *The Theory of Social Economy*, translated by S.L. Barron, New revised edition, New York, Augustus M. Kelley Publishers, (coll. Reprints of Economics Classics, VIII+689 pages.): 11 et seq.

Harrod, R. F. (1931). The law of decreasing costs, *Economic Journal*, vol. 41, No. 164, December: 566–576.

Marshall, A. (1890). *Handbuch der Volkswirtschaftslehre* (published in English as: *Principles of Economics*), vol. 1, Book V.

Pareto, V. (1906). *Manuel d'économie politique*, (published in English as: *Manual of Political Economy*), Paris 1927: 148, No. 10.

Peiser, H. (1924). *Einfluss des Beschäftigungsgrades auf die industrielle Kostenentwicklung* (Influence of the Employment Level on Industrial Cost Trends). Berlin, 10.

Schmalenbach, E. (1927). see, *Der Kontenrahmen* (published in English as: *Framework of Accounts*). Leipzig, 31.

Schmalenbach, E. (1928). *Wiener Rede* (Vienna Lecture), Summer.

Schmalenbach, E. (1930). *Grundlagen der Selbstkostenrechnung und Preispolitik* (published in English as: *Foundations of Cost Accounting and Pricing Policy*), 5th German ed., Leipzig 1930: 32 et seq.

Schmalenbach, E. (1926). "Abschreibungen" (Depreciation and Amortisation) in *Grundlagen dynamischer Bilanzlehre* (published in English as: *Dynamic Accounting and Price-Level Adjustments: An Economic Consequences Explanation*), 3rd German ed.: 104 and 113.

Chapter 2
Costs in Single Production

§ 1. The Fundamental Concepts of Cost Theory

I.

We have already been introduced to the term "total costs" in the previous chapter and we know that total costs are a definite, monotonically increasing function of the velocity of production. We would like to study this function a little more closely.

Let us imagine that the firm starts to increase its velocity of production. Then this firm's total costs also increase. But how much do they increase by? Here we must note that the firm only increases its velocity of production by increasing direct input. Let us assume that the quantity of indirect factors of production stays the same. Usually this must lead to the firm becoming relatively more unproductive for any velocity of production, that is, only a decreasing increase in products can be obtained by the same total cost increase with an increasing velocity of production. This consequence is not proven conclusively by the invariability of the indirect factors of production, but it is very convincing. It is obvious if we consider the following: indirect factors of production are a necessary condition for the production of products. Where they remain unchanged, the composition ratio of the input vector components changes with the increasing velocity of production to the disadvantage of the indirect factors of production. It is to be expected that this circumstance has an impact on the firm's productivity as described above. This fact is particularly apparent in agriculture. If we regard land as being an invariable factor of production and compare it with the other factors of production which we regard as being variable (by extending our focus over a correspondingly long period of time), then the "law of diminishing returns of the soil" (Translator's note: Brinkmann English ed. 1935:7) or as Brinkmann[1] quite rightly describes it, the "law of diminishing rate of increase in return" (Brinkmann English ed.: 7) very clearly results.

---

[1] Brinkmann, Th. (1935): 7.

H. von Stackelberg, *Foundations of a Pure Cost Theory*,
DOI 10.1007/978-3-642-34537-1_2, © Springer-Verlag Berlin Heidelberg 2014

This law is simply a special case of the circumstances we formulated above. If it did not apply, then we could produce any arbitrary product quantity with a sufficient input of the variable factors of production without seeing any rise in the increase in costs (Translator's note: "increase in costs" see Peacock: 47) for a limited plot of land or even for an arbitrarily small plot, which is generally agreed to be impossible[2].

What happens when the different production levels and the cheapest input vectors required for that purpose are considered in general terms, irrespective of an actual firm, that is, assuming all factors of production to be arbitrarily variable? Is it also the case here that, starting from any point, total costs increase proportionately more quickly than the velocity of production where the velocity of production is increasing? We cannot prove this proposition so convincingly in the same way here. However, it follows almost as a necessity from another fact. This fact is the principle of scarcity. Experience tells us that it is impossible to arbitrarily obtain large velocities of production with a fixed quantity of factors of production. Then however, growth in the velocity of production requires an increase in the factors of production, but in the market economy these factors are limited in accordance with the principle of scarcity. Hence, the factors of production cannot firstly be increased by an arbitrary amount and secondly they can only be increased from any point with increasing factors of production prices, that is, when there is an increasing rise in costs[3]. In general terms, this results in the law stated above.

The difference between the first and the second reason for this law is the fact that the velocity of production which is generated by the increase in the total cost increase, is as a general rule much lower in the first case than in the second. The second case naturally also applies to the individual enterprise but it may be that now the velocity of production in question is so high that it is not achieved for economic reasons. Conversely, in the first case, the velocity of production is relatively low at the point where total costs start to increase.

Yet another feature of the total cost function is revealed from the first point of view. In an identical fashion to high velocities of production, a firm's factors of production are composed of relatively few indirect factors of production, so with low velocities of production there are either relatively large amounts of the indirect factors of production or relatively small amounts of the direct factors of production. Hence the composition ratio of the components of the input vector has shifted to the detriment of the direct factors of production. Where the velocity of production increases, the composition ratio of the factors of production is lower at first, that is, the increase in costs decreases.

From this we can derive the following rule (which is not generally applicable) for the total cost function. If we let the velocity of production increase from 0, total costs increase continually. But they firstly increase in decreasing amounts, that is,

---

[2] William Stanley (1871): 200. Also: Barone, E. (1927): at the end of § 10 and § 11.

[3] See Barone, E. ibid.: § 9 and § 10. By contrast: see Bücher, K. (1921a): 92, footnote.

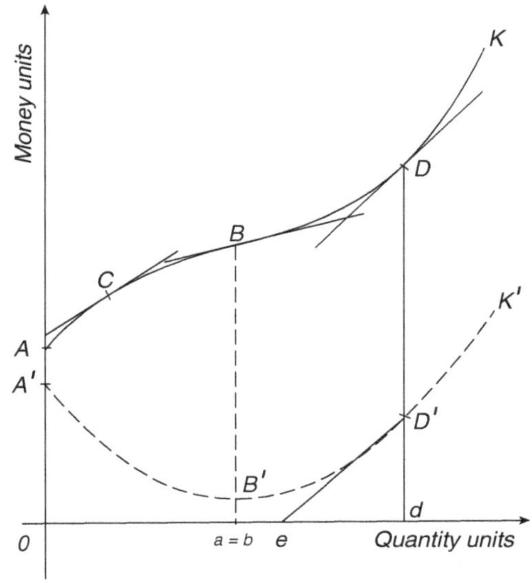

Fig. 2.1 A regular and smooth total cost function

the increase in total costs initially decreases until the velocity of production has achieved a certain level. If the velocity of production continues to increase, an increase in the total cost increase begins from any point and this is perhaps further increased by another increase in the velocity of production[4].

We can see this rule for the total cost function appearing in Fig. 1.1 of Chap. 1. The total cost curve is concave downwards between the origin and point *A* and is subject to the law of increasing returns. The law of diminishing returns prevails from *A* onwards. If we still assume (which we will continue to assume from now on, provided nothing else is specifically stated) that the total cost function is regular and smooth, that is, it is continuous and can be repeatedly continuously differentiated, we can then illustrate the shape of this function with the help of Fig. 2.1[5].

The rule described above is not generally applicable. In addition, we therefore also have to consider total cost functions that have a different shape, especially functions for which the increase in costs continuously decreases, as well as functions for which the increase in costs remains invariable. The importance of this observation is only increased by the fact that the regular total cost function also shows a diminishing rate of increase in costs when the velocity of production is low. In order to remain faithful to generally used wording, in the following pages we would like to express the fact that the increase in costs increases (remains constant,

[4] For this total cost function rule, see Barone, E. ibid. § 8 to § 13. Also: Kalischer, H.E. (April 1929 and especially January 1930): 18 et seq.

[5] Schneider, E. provides another rule. See Schneider, E. (1931a): 585 et seq. Also see German ed. pp. 590 in the same cited literature.

decreases) using the following proposition: "The enterprise is subject to the law of diminishing (or constant or increasing) returns". In terms of its wording, this proposition does indeed set out what we are dealing with here, but only very incompletely, as Brinkmann (see above) correctly notes[6]. But its typical application shows that it is considered to be a fact and we want to express this now. Using this terminology, the description of the regular total cost function would read as follows: The enterprise is subject to the law of increasing returns for lower velocities of production. If the velocity of production increases above a certain level, the enterprise is then subject to the law of diminishing returns. There is a line segment or even just one point between these levels where the law of constant returns applies[7].

II.

In our observation a moment ago we repeatedly used the expression, "increase in costs". We would now like to concentrate more closely on this variable.

The increase in costs is the change in total costs which occurs if we either increase or decrease the velocity of production by any amount. Depending how this amount is chosen and whether we either add it or take it away, the (positive or negative) increase in costs will then be different. We now wish to introduce a measure for the increase in costs. This is the increase in costs translated into the unit of the velocity of production. Where there is an increase in costs, we obtain the level of the increase using the corresponding change in the velocity of production[8].

---

[6] Also see Barone, E. ibid. § 11, last paragraph.

[7] See Schneider, E. for these explanations and the cost categories defined in the following pages, 269 et seq., sections A and B. It is not possible to agree with part of Section C of Schneider's article. In particular, the explanations which lead to this proposition on pp. 292 of the German ed., "The divergences of the actual average cost curve (Translator's note: "average cost curve" see Peacock, A.: 187) from the planned average cost curve depend exclusively on the size of the share of fixed costs..." might be untenable, as well as the following propositions. Parabola Ψ (x) in equation (18), has a horizontal tangent at its minimum position ($x = x_0$). Therefore it is intersected at point Ψ (x_0) by each sloping curve which goes through this point. To the right or left of this point, the curve then follows a path above parabola Ψ (x_0). Therefore Schneider's condition (16) for zero a is able to be fulfilled within a certain interval that is limited at one end by x_0. As a result, a lower limit for α cannot be determined in the form given by Schneider.

[8] We therefore obtain the variable that Schmalenbach, E. originally termed the "proportional cost" (Translator's note: see Peacock, A.: 231) and which he now describes as "marginal costs" (see *Selbstkostenrechnung*, German ed. pp. 52). We do not consider this name change to be a very fortunate one. Economic theory keeps the term "marginal costs" (Grenzkosten) to be used for the differential quotient of the total cost function. However, this now becomes a question of a differential quotient which is an approximate value of the differential quotient. It is important to provide different descriptions for the value that is to be approximated and the value that does the approximating.

If the total cost function is regular and smooth (which is what we shall assume), the more the different amounts of the increase in costs associated with a velocity of production equal each other, the smaller the changes in the velocity of production are assumed to be. If the changes are allowed to become increasingly smaller, all corresponding amounts of the increase in costs then strive for a single value. This value may be defined as the amount of the increase in costs in the immediate vicinity of a velocity of production. This is merely the first differential quotient of the total cost function.

For each velocity of production, there is a number which represents the amount of the increase in costs. Its size is hence a function of the velocity of production. In the following pages we would like to describe it using the rather unfortunate term, "marginal cost level" or even simply "marginal costs", in line with everyday language. Since the marginal cost level is the first differential quotient of the total cost function we would like to introduce the symbol K' for the marginal cost level[9].

Everything we said in section I about the increase in costs also applies to the marginal cost level. Because this term is precise, it is the only term we will use from now on.

III.

In choosing the term for the first derivative of the total cost function, we would usually be able to refer to it as the "total cost gradient" since this derivative is just the measure of the gradient of the basic function at a particular point. We have reserved this term for the derivative of the marginal cost level. We would like to describe this derivative as the "marginal cost gradient". It is simply a measure of the gradient (which can be either positive or negative) of the increase in costs. Wherever we referred to the gradient of the increase in costs in section I, we can now substitute the term "marginal cost gradient". Since the marginal cost gradient of the second differential quotient is the total cost function, we can hence insert the symbol K''. K'' is also a function of the velocity of production[10].

Using precise mathematical terminology we can describe the regular total cost function in the following way:

---

[9] It therefore follows that:

$$K' = \frac{d\,K\,(x)}{d\,x} = K'\,(x).$$

See Amoroso, L. ibid.: 4, "il costo marginale" (marginal costs).

[10] We therefore derive the following:

$$K'' = \frac{d^2\,K\,(x)}{d\,x^2} = K''\,(x).$$

1. The law of increasing returns, $K'' < 0$, applies to all x below a particular variable a, that is, in the interval $(0, a)$.
2. The law of constant returns, $K'' = 0$, applies above a up to a particular variable b, that is, in the interval (a, b).
   Where the total cost curve is regular and smooth, i.e. it can be repeatedly, continuously differentiated, then the two variables a and b decrease. Then K'' has a zero value only when $x = a = b$.
3. The law of diminishing returns applies above b, that is, for all $x > b$: $K'' > 0$.

Since K increases monotonically, K' is positive everywhere. And indeed K' declines in the interval $(0, a)$, is constant in the interval (a, b) and increases for all $x > b$.

Figure 2.1 illustrates this situation.

Total cost curve K is concave downwards from point A to point B. The enterprise is hence subject to the law of increasing returns for all velocities of production between point 0 and point b. The marginal costs that are fixed at every point on the curve by the tangent of the directional angle to the tangent curve at that point, clearly decrease between 0 and b and when the relevant point on the curve moves towards the right, the velocity of production increases. We can see that, for example, the tangent at point C is steeper than the tangent at point B (B lies further to the right).

The enterprise is subject to the law of diminishing returns for all points lying to the right of B, that is, for all velocities of production which are greater than $\overline{Ob}$. We can immediately see that the tangent at point D is steeper than the tangent at point B (B lies further to the left).

Point B hence stands out as the curve's point of inflexion. All points on the curve to the left of point B lie below the inflexion tangent and also below each tangent in the curve portion $\widehat{AB}$. All points on the curve to the right of B lie above the inflexion tangent and also above each of the other tangents to the right of B.

Since marginal costs are fixed by the directional tangent of the total cost curve at every point, they can be constructed from the total cost curve. In Fig. 2.1 they are constructed for point D on the total cost curve with base point d and hence for velocity of production $\overline{Od}$. We move the unit of length away from d to the left. Let $\overline{ed} = 1$. Using e, we pull the parallel to the tangent towards the total cost curve at point D. This parallel intersects the ordinate $\overline{dD}$ of point d at D'. Then $\overline{dD'}$ is the required marginal cost level and D' is therefore the point on the marginal cost curve that corresponds to the abscissa $\overline{Od}$. It is in fact $\sphericalangle\, deD'$ of the directional angle of the tangent at D. Its tangent is $\frac{\overline{dD'}}{\overline{ed}} = \overline{dD'}$ because it follows that $\overline{ed} = 1$.

Using point construction we arrive at marginal cost curve K$'$ shown by a dashed line in Fig. 2.1. Its lowest point, the point where marginal costs no longer decrease, is B$'$.

The marginal cost increase curve is derived from the marginal cost curve in the same way that the marginal cost curve is derived from the total cost curve. We have not drawn it in separately.

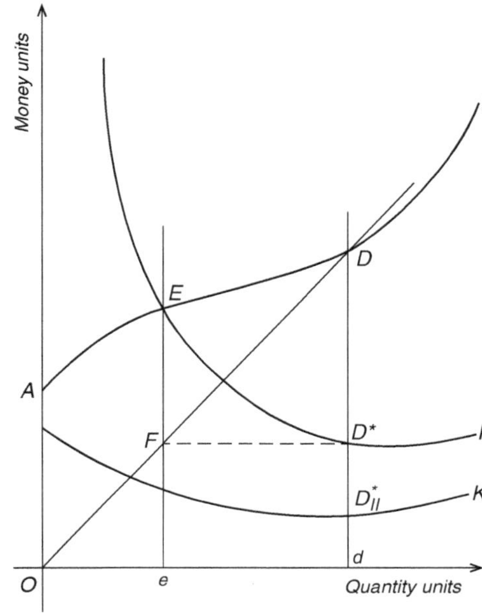

Fig. 2.2 Average cost and average variable cost functions

IV.

Finally, we still need to introduce a function which we have not seen yet but which we will need later. This is the average cost function. A certain velocity of production requires certain total costs. These total costs, divided by the associated velocity of production, give average costs. We introduce the symbol K^* for this new function. K^* is also a function of the velocity of production[11].

Similarly, we can use the term "average variable costs" to indicate when we have divided the total costs associated with a velocity of production, not by the total costs, but instead by the variable costs. As with K^*, we describe this new variable using K^*_{II} (see footnote 11).

K^* and K^*_{II} are shown in the graph in Fig. 2.2.

[11] We hence derive:

$$K^* = \frac{K(x)}{x} = K^*(x)$$

$$K^*_{\text{II}} = \frac{K_{\text{II}}(x)}{x} = K^*_{\text{II}}(x)$$

$$K^* = \frac{K_{\text{I}}}{x} + K^*_{\text{II}}.$$

With increasing x, the two functions $K^*(x)$ and $K^*_{\text{II}}(x)$ become closer to each other asymptotically because K_{I} is constant and $\lim\limits_{x\to\infty} \frac{K_{\text{I}}}{x} \to 0$.

Average costs are defined as the quotient of the ordinate and abscissa of a point on the total cost curve. If we describe the connection between the origin and a point on the curve as the position vector of this point on the curve, we can hence say that average costs are defined as the tangent of the angle between the quantity axis and the position vector of the total cost curve. The construction of the point on the curve of the average cost curve also results from this definition. It is constructed in Fig. 2.2 for velocity of production d. At unit point e ($\overline{Oe} = 1$), we constructed perpendicular $\overline{eE}$ and we projected the intersection point F of perpendicular $\overline{eE}$ and position vector $\overline{OD}$ onto the ordinate $\overline{dD}$ of the relevant point on the total cost curve at D. Projection D^* is the required point of average cost curve K^*, whose base point is d. These are in fact the average costs of the velocity of production $\overline{Od}$ of quotient $\frac{\overline{dD}}{\overline{OD}}$.

Using intersection theory, it follows that:

$$\frac{\overline{dD}}{\overline{\overline{Od}}} = \frac{\overline{eF}}{\overline{\overline{Oe}}} = \overline{eF} = \overline{dD^*},$$

and that $\overline{Oe} = 1$ and $\overline{eF} = \overline{dD^*}$ using construction.

Hence we have constructed the average cost point on the curve that corresponds to velocity of production $\overline{Od}$. By constructing additional points, the average cost curve K^* eventually emerges as the geometric position of the points. K^*_{II} occurs using a corresponding construction by firstly setting point A instead of O and afterwards shifting the curve that is found by $\overline{OA}$ ($= K_1$) downwards.

With the help of the functions defined and described in this section, we will continue our analysis of the regulatory laws of production in the following pages. We will firstly focus on velocities of production, which are important from the firm's internal perspective, in order to then study the enterprise's situation in the market.

§ 2. The Optimum Position

I.

The enterprise can achieve various velocities of production with various levels of total cost. We now wish to investigate which velocity of production is the lowest in relative terms. We would firstly like to set out what has to be understood in greater detail. The price that must be paid for the unit of a quantity of products produced in the unit time, and hence the corresponding total costs that must be exactly covered by returns, is equal to average costs. This is because average costs multiplied by the velocity of production (that is, multiplied by the number of units of product (Translator's note: "units of product" see Cassel: Book 2, Chapter VII, § 29) produced in the unit time), exactly equate to total costs (ex definitione). We shall call this price the "cost covering price". Every velocity of production has a corresponding cost covering price. The velocity of production with the lowest cost covering price is clearly the lowest because the cost-covering price of this velocity of production is the lowest of all the prices at which the enterprise can sell without making a loss. The enterprise can also sell at this price without making a loss, but only at precisely this velocity of production and at no other. Therefore we shall call this the optimum velocity of production. We shall describe the enterprise's general situation if it achieves the optimum velocity of production as the optimum position (Translator's note: "optimum position" see Peacock: 50, often referred to in modern parlance as the 'Minimum Efficient Scale'. Although von Stackelberg uses 'Betriebsoptimum' in German, literally 'firm's optimum', he seems to apply this term to firms as well as enterprises.) However, it would not be completely correct to assume the enterprise would always have to achieve its optimum position. Usually the velocity of production which must be achieved by the enterprise will differ even to the optimum based on the law which we will explain later. For the enterprise, the optimum position is simply an excellent position arising from certain features. Our task is now to determine this optimum position more precisely.

II.

Which velocity of production is the optimum? According to the definition, it is the velocity whose cost covering price is the lowest because this price is equal to average costs and so the optimum velocity of production is characterised by having the lowest average costs. Expressed another way, the optimum velocity of production has the minimum level of the average cost function as its average cost.

Figure 2.3 serves to illustrate this situation for us.

Since average costs are defined by the directional tangent of the total cost position vector, to determine the optimum position we must look for the flattest position vector.

This is clearly the position vector which is formed in such a way that no point on the curve lies between it and the quantity axis. In fact, if a point is lower in that area, its position vector will hence be lower.

If the curve has a tangent at the optimum point P, this is then identical to the position vector. It can be expressed another way: the optimum position tangent is characterised by the fact that it passes through the origin. This means however:

(I). At the optimum position, marginal costs and average costs are equal to one another.

We describe this proposition as the fundamental proposition (Translator's note: "fundamental proposition" see Baumgärtner, (2001): 514) of the optimum position.

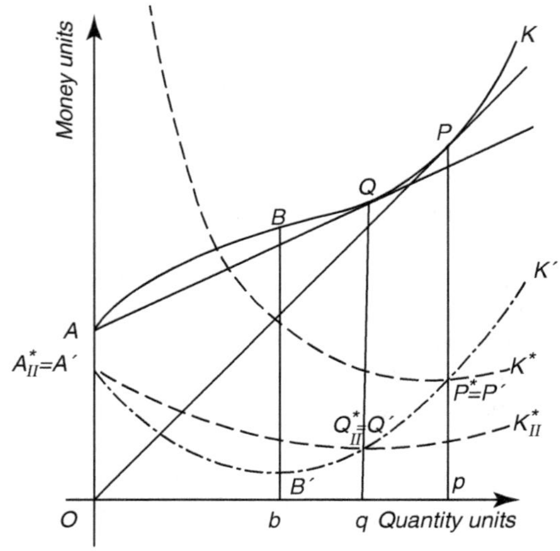

Fig. 2.3 The optimum velocity of production

It shows a surprising feature of the optimum position. This can therefore also be defined by the intersection point of the marginal cost curve and the average cost curve. In Fig. 2.3, $\overline{Op}$ is the optimum velocity of production and $\overline{pP}$ is the cost covering price.

III.

1. Another important feature of the optimum position is the fact that the total cost curve is convex downwards at this point (see Fig. 2.3). However if the total cost curve was concave downwards at this point, there would be position vectors lower than position vector $\overline{OP}$. Then P would not be the optimum position. We know however that where the total cost curve is convex downwards, the law of diminishing returns applies[12]. As a result, the following proposition applies:

(II). The law of diminishing returns applies to the optimum position.

This also means that the optimum velocity of production must always be higher than variable b defined in (1) above. So long as an enterprise is subject to the law of either increasing or constant returns, it cannot possess an optimum position. The relatively lowest velocity of production therefore does not lie where the increase in costs is lowest, but instead goes beyond this point by a considerable amount[13].

[12] See § 1, III of this chapter.

[13] This amount may be an estimate. See Appendix A.

2. An important situation occurs if we compare the marginal costs and average costs for all values of x. The following proposition in fact results:

(III). Where the total cost curve is regular and smooth as well as regular (Translator's note: the repetition of "regular" in this English sentence occurs as a result of translating the two German terms "regulär" and "regelmäßig" using "regular and smooth" and "regular" respectively), average costs for all velocities of production are higher than marginal costs below the optimum, whereas they are lower for all velocities of production above the optimum.

This proposition for the regular and smooth total cost curve may be clearly proven using analytical methods. The proposition is very convincing when looking at Fig 2.3. All tangents to the regular total cost curve between points A and P in fact meet the ordinate axis at its positive section and are hence not as steep as the corresponding position vector. All tangents to the right of P by contrast meet the ordinate axis at its negative section and are hence steeper than the corresponding position vector. So, the ordinates of marginal cost curve K' are also smaller between O and p, whereas to the right of p they are larger than the corresponding ordinates of average cost curve K*.

As soon as the total cost function is known, this proposition allows us to recognise immediately whether each velocity of production is either higher or lower than the optimum velocity. Where average costs are higher than marginal costs, the velocity of production is lower than the optimum, and average costs are decreasing. Where average costs are lower than marginal costs, the velocity of production is higher than the optimum and then average costs are increasing.

We would now like to present two terms which Schmalenbach[14] introduced; they are very helpful in describing the situation the enterprise currently finds itself in. These are terms which are not tied to the regular total cost function, but can be applied generally. Where average costs are higher than marginal costs, we shall therefore say that total costs are degressive and here the enterprise finds itself in cost degression (Translator's note: "cost degression" see Peacock: 51). Where average costs are lower than marginal costs, we shall then say that total costs are progressive. Here, the enterprise finds itself in cost progression (Translator's note: "cost progression" see Peacock: 296).

We will show the extent to which these terms are used in the same sense Schmalenbach used them later in this book[15].

With these terms, our proposition is as follows:

(IIIa). Cost degression exists below the optimum position while cost progression exists above this optimum.

If we consider that as the cost curve changes, the optimum velocity of production becomes increasingly large, then the area that is subject to cost degression also becomes increasingly large. We can also formulate the above proposition as

[14] Schmalenbach, E., *Selbstkostenrechnung* (Cost Accounting): 32 et seq.

[15] See Appendix C.

follows. As long as the optimum position has not yet been achieved, cost degression exists. It follows that:

(IIIb). Enterprises to which the law of increasing returns applies, are subject to cost degression[16]. The same also applies to the case of constant returns.

On the other hand, it is important to establish that if the enterprise is subject to the law of diminishing returns for all velocities of production, these velocities of production nevertheless show cost degression in a particular opening interval. The larger this interval is (ceteris paribus), the larger K_{I} is. In the following sections we will see that p then only coincides with the origin under the conditions assumed here if constant costs have a *0* value.

§ 3. The Minimum Position

I.

Let us now turn to another problem, which is very similar to the previous one, where our starting point was to determine the lowest cost covering price and its corresponding velocity of production. Now let us ask ourselves what the lowest price is at which the enterprise could still continue production without incurring a greater loss than if it ceased production (for a short time). This lowest price does not completely coincide with the lowest cost covering price. From this we can make the following observation. Constant costs are the amount the enterprise must bear under any circumstances, even if the firm lies idle. The greatest loss that the enterprise can incur during active production, without being in a less favourable position than if it lies idle, is therefore equal to constant costs. The price we are looking for here thus only needs to cover variable costs. It is hence equal to average variable costs and, based on a similar observation to the one stated at the start of the previous section, because we are looking for the lowest of these prices we must determine the minimum level of average variable costs. The velocity of production which minimises average variable costs we term the minimum and we term the enterprise's corresponding position as the 'minimum position' (Translator's note: "minimum position" see Peacock: 50, often referred to in modern parlance as 'Minimum Average Variable Costs'. Although von Stackelberg uses 'Betriebsminimum' in German, literally 'firm's minimum', he seems to apply this term to firms as well as enterprises.)

The minimum position corresponds to the optimum position when constant costs are zero. Principles follow from this conclusion and they are derived in exactly the same way as the corresponding propositions for the optimum position. The minimum position is shown on the graph by constructing the flattest line from point *A* to a point on the curve and then leaning the tangent against the total cost curve from

[16] Because in fact the whole cost curve is concave downwards, hence $p \text{ - } \rightarrow \infty$.

point A. Q represents the point on the curve which is the minimum position and q is its abscissa. Figure 2.3 indicates that the features of Q differ from the features of P only in terms of the parts which depend on constant costs. The geometric position results from its similarity to the optimum position based on Fig. 2.3.

We firstly obtain the fundamental proposition for the minimum position:

(IV). In the minimum position, marginal costs and average variable costs are equal to each other[17].

As a result, the minimal velocity of production is revealed as being constructed from the abscissa of intersection point $Q^*_{II} = Q'$ of the average variable cost curve and the marginal cost curve.

Furthermore it is easy to see that point Q, just like point P, lies on the convex branch of the regular total cost curve, and is in fact between B and P (see proposition VIII below).

(V). The law of diminishing returns then also applies to the minimum position.

Consequently, $q > b$. If an enterprise is subject to the law of increasing returns, it cannot have an minimum position. However it is easy to see that unlike the optimum position, this does not always apply to the law of constant returns. In particular, there are marginal cases where b and q coincide. For example, where we have an enterprise that is subject to the law of constant returns for an opening interval and then subject to the law of diminishing returns[18], every point between O and b can therefore be regarded as the minimum position because the flattest line from point A to the total cost curve now coincides with the first curve portion.

It is different if the enterprise is subject to the law of diminishing returns for all velocities of production. Here b and q coincide with the origin. The flatter the position vector for A is to the total cost curve, the shorter it is and the closer this point on the total cost curve then lies to point A. The same applies to the tangent to this total cost curve.

From Fig. 2.3 we can immediately see that the marginal cost curve and the variable average cost curve meet the ordinate axis at one and the same point A. All regular and smooth[19] total cost functions have this feature. The position vector for

---

[17] See also Amoroso, L. ibid.: 5: His "punto di fuga" (vanishing or lowest point) is merely our minimum position which can also be very easily seen from Fig. 1 of his article. By contrast, his Fig. 2 (Italian ed. pp. 7) achieves the same circumstances as our "optimum". Here, his "prezzo di fuga" (vanishing price) which he calculates using $b + 2\sqrt{a\,c'}$, is in fact our optimum price. He contradicts the definition of the features Amoroso, L. gives to the "prezzo di fuga" in Italian ed. pp. 5 of his article.

[18] See Schneider, E. ibid.: 22, footnote 2).

[19] We should particularly draw the reader's attention to the fact that in our context "regular and smooth" and "regular" (Translator's note: "regulär" and "regelmässig" respectively in German) are used as two completely different terms. "Regular" ("regelmässig" in German) means "in accordance with the rule" whereas "regular and smooth" ("regulär" in German) means, "can be repeatedly, continuously differentiated", that is, "smooth" ("glatt" in German) (without corners) for K, K' and K''.

A at point R on the total cost curve in fact coincides with the tangent to the total cost curve at R if point R moves to A. We can hence say:

(VI). The lower the velocity of production the less its marginal costs and variable average costs differ from each other[20].

II.

The following proposition applies to the minimum position in the same way it does to the optimum position:

(VII). Where the total cost curve is regular and smooth, as well as regular (Translator's note: please see our earlier remarks about the necessity of repeating "regular" in the English translation), then average variable costs for all velocities of production below the minimum are higher than marginal costs, although lower for all velocities of production above the minimum.

This proposition becomes clear in Fig. 2.3.

All tangents to the regular cost curve between points A and Q in fact meet the ordinate axis above A and are hence not as steep as the intersection lines for A after the corresponding points on the curve. Conversely, all tangents to the right of Q meet the ordinate axis below A and are hence steeper than the corresponding intersections for A after the points on the curve. Correspondingly, the ordinates for K' are also smaller to the left of Q^*_{II} and larger to the right of Q^*_{II} than the corresponding ordinates for K^*_{II}.

This proposition allows us to establish whether an arbitrary velocity of production is either higher or lower than the minimum velocity. In the first case, $K'(x) > K^*_{\mathrm{II}}(x)$ results, while in the second, it is $K'(x) < K^*_{\mathrm{II}}(x)$.

We can also apply the terms "progressive" and "degressive" to variable costs. Then the following proposition applies:

(VIIa). Variable costs are degressive below the minimum position and progressive above it.

We shall now determine the location of point q compared to point p. Both points lie above point b, above which the marginal cost level increases. Function $K'(x)$ is now monotonically increasing.

$K'(x)$ has the value $K^*_{\mathrm{II}}(q)$ for $x = q$ and the value $K^*(p)$ for $x = p$.

Now $K^*_{\mathrm{II}}(q) < K^*_{\mathrm{II}}(p) < K^*(p)$. This leads to $K'(q) < K'(p)$ and therefore $q < p$ because $K'(x)$ increases monotonically. Consequently, q lies between b and p[21].

[20] We can also see this situation achieved in Fig. 1 of the article cited by Amoroso, L. (original, non-English ed. pp. 5), whereby it is clearly evident that Amoroso, L. is discussing the minimum position. This element is missing from his Fig. 1 (original, non-English ed. pp. 7). Here this is simply a matter of the optimum position.

[21] See the observations about proposition (IV).

We can hence put forward the following proposition:

(VIII). Total costs are degressive between the minimum position and the optimum position, whereas variable costs are progressive.

Let us imagine (by changing the total cost function) that the minimum velocity of production becomes increasingly larger and so the area in which variable costs are degressive will also become increasingly larger. From this, in exactly the same way as for the optimum position, it follows that:

(VIIb). This law of increasing returns applies to enterprises subject to variable cost degression.

By contrast, the law of diminishing returns applies to enterprises for all velocities of production, whereas enterprises which have no velocity of production are subject to variable cost degression, as we have already seen above (proposition V, second clause).

IV.

To summarise, we can establish the following features of regular total cost functions (see Fig. 2.3).

1. Up to point *b*, marginal costs decrease, after which they increase.
2. Up to point *q*, average variable costs decrease, after which they increase. At *q* they are equal to marginal costs, below it they are higher than marginal costs and above it they are lower. Below *q*, variable costs are in degression but above, they are in progression.
3. Up to point *p*, average costs decrease, after which they increase. At *p* they are equal to marginal costs, below *p* they are higher than marginal costs and above it they are lower. Below *p* the enterprise is in cost degression but above, it is in progression.
4. The ranking of the points we have mentioned is: *O, b, q, p*. Potentially, point *a* may still be inserted between *O* and *b*. Total costs would then be linear between *a* and *b*.
5. The following model gives us an overview of the behaviour of the four functions, namely, total costs, marginal costs, average variable costs and average costs, in individual sections of the velocity of production scale:

Intervals on the velocities of production scale	Decreasing	Increasing
(O, b)	K'; K^*_{II}; K^*	K
(b, q)	K^*_{II}; K^*	K; K'
(q, p)	K^*	K; K'; K^*_{II}
(p, ∞)	—	K; K'; K^*_{II}; K^*

All these relationships can be seen in Fig. 2.3 if we look at the individual curves K, K', K^*, K^*_{II} [22].

V.

The observations we made in the last two paragraphs initially apply only to "short periods of time" according to Alfred Marshall (Translator's note: see Book V, Chapter V: 363). However Marshall has shown[23] that relationships over long periods operate subject to the same laws as short periods. The only difference is that the longer the period of time is over which we choose to make our observation, the smaller the share of total costs that can be described as constant. Alfred Marshall explained this fact at great length in Book Five of his major work. We therefore believe that we can do without describing it now and move straight on to the consequences arising in the course of our investigations from the difference between long and short periods of time. From now on we shall describe these long and short periods of time as "Marshall periods of time", in line with Alfred Marshall's meaning and usage (Translator's note: see Marshall (1961): 342). With the help of terminology now available to us, we can formulate the different appearance of the enterprise depending on the length of the Marshall period of time in the following way:

[22] Strangely, there is no agreement between individual authors in their definitions of the terms "increasing returns" and "diminishing returns" which are used so often. One group describes the situation in which marginal costs decrease as "increasing returns", that is, $K'' < 0$ and describes the opposite situation ($K'' > 0$) as "diminishing returns". The following authors belong to this group, e.g.: Ricardo, D. (1921): 52 et seq, Jevons, W.S. (1924): 198 et seq., Marshall, A. (ibid.: 188–209), Pareto, V. (1897): 102 et seq., Cassel, G. (1918): 252 et seq., Brinkmann, Th. (ibid.). We have also aligned ourselves with this terminology. By contrast, other authors use the terms "increasing returns" synonymously with "degressive costs" and use "diminishing returns" synonymously with "progressive costs", for example, Barone, E. (ibid, 10/11) and Bowley, A. (ibid.: 33 et seq.). This second term then builds on the decrease or increase in average costs (possibly also, in average variable costs). The difference can easily be seen from our explanations. Edgeworth, F.Y. has already examined this in great depth in "The Laws of Increasing and Diminishing Returns", Papers...I: 61 et seq.. A very similar observation was presented by Pigou, C.A. (1927): 188 et seq. Close attention must be paid to this diverse terminology because it can give rise to many misunderstandings.

[23] Alfred Marshall, ibid.: Book V.

(IX). The longer the Marshall period of time is, the closer the minimum position and the optimum position become.

This proposition can be seen in the following way: the longer the Marshall period of time is, the closer the optimum position moves towards the point at which the relatively cheapest production of all occurs, using the production possibilities that exist in the branch of production concerned. The same also applies however to the minimum position because the longer the Marshall period of time is, the lower constant costs are as a proportion of total costs and hence the least different the average variable cost minimum position is from the average cost minimum position. However, because these two values are both marginal cost function values and are in fact in the section of the curve where the value is already increasing monotonically, as these values move closer, their abscissae and hence their corresponding velocities of production move closer to each other too.

§ 4. The Enterprise's Supply According to the Profit-Making Principle

In the first three sections of this chapter we described the enterprise with regard to the organisation of its costs. We looked at the enterprise as a demanding and producing element of the market economy, rather than a supplying element. Our propositions arose from the notion of costs, the scarcity principle and the economic principle. We were therefore able to establish the situation an enterprise finds itself in for a given production level. However, we have not concerned ourselves with the question of how this production level comes about. In order to determine this, we must regard the enterprise as a supplier. And indeed we must now use either two other principles or else groups of principles, that is, we must determine the motivation for production as well as its market situation. In this section, as the title already indicates, we apply the profit-making principle as the enterprise's motivating principle. In the next section we will consider the enterprise when the principle of the satisfaction of needs and wants applies.

I.

1. We defined the profit-making principle as striving to obtain the highest possible profit via production. By profit, we mean the difference between returns and total costs, where the return[24] is the enterprise's revenue from selling its products produced in unit time. Returns are hence the product of price and velocity of production so long as the goods produced are completely sold. From now on, we

[24] See also chap. 1, § 2, III.

shall assume that the quantity produced and the quantity sold are always identical. We should be able to do this and for the following reason: what we would like to investigate is how to determine production using the market situation. However, determining production does not occur immediately, but instead develops through an awareness shown by the entrepreneur. And indeed, it is not the market situation itself that is key for production, but instead it is the picture the entrepreneur has of the market situation. Then however it is a reasonable assumption that the entrepreneur always produces as much as he believes he can sell.

We would like to introduce the symbol E for returns. "Returns" are then defined according to the definition:

$$E = x \,.\, P.$$

Returns are hence a function of x and are furthermore a function of the variables whose function is P. In any case, returns are a function of the velocity of production. They of course also depend on countless other variables. However, we do not need to go into more detail now. We are only interested here in the economic dependence of returns on those variables that depend on the entrepreneur's intentions. However, we will now simply look at the velocity of production. The price can (in the case of a monopoly) actually also depend on the entrepreneur's intentions. However, the entrepreneur cannot determine the velocity of production and price independently of each another. At a selected price, the entrepreneur can only sell either a certain quantity or conversely, where the quantity is selected first, he can only sell at a certain price that is no longer dependent on his arbitrary choice. In other words, the entrepreneur can only influence returns by economic methods using the velocity of production. We have thus had to consider returns to be a function of the velocity of production[25]. For all other variables that are outside the scope of the enterprise, we will always state, "ceteris paribus".

2. The fundamental issue for production, according to the profit-making principle, is what velocity of production must be achieved for a given market situation in order to obtain maximum profit[26]?

For profit, we insert the symbol G. Then according to the definition, the following applies:

$$G = E\,(x) - K\,(x) = G\,(x).$$

Profit appears to be a function of the velocity of production. Each velocity of production corresponds to a particular profit (which of course can also be negative. Its absolute amount is the loss made by the enterprise). We then ask, "What velocity of production maximises profit?"

---

[25] It follows that: $E = E\,(x)$.

[26] For the following, see particularly Cournot, A. (1924).

The answer emerges from a simple idea. First of all we introduce the symbol s for the required velocity of production that we shall describe as the most favourable velocity. This most favourable velocity of production s stands out because every other velocity of production results in a smaller profit. In other words, profit[27] increases with an increasing velocity of production until this has achieved the value s; then it decreases. However, with an increasing velocity of production, total costs $K(x)$ increase. Profit hence increases if returns increase more quickly than total costs, but it decreases if returns increase more slowly than total costs. The most favourable velocity of production s is now characterised by the fact that the increase in returns and the increase in total costs equal each other.

Let us describe the size of the increase in returns as the marginal return, in a similar way to that of our cost terminology, remaining consistent with the wording generally used. We hence arrive at the fundamental proposition of the profit-making principle:

(X). Marginal costs and marginal returns of the most favourable velocity of production are equal to each other.

We describe the marginal return, which is just the first differential quotient of the revenue function, using $E'(x)$. Then our fundamental proposition is repeated using the equation $E'(s) = K'(s)$.

Before we go any further, we would like to illustrate this proposition in the graph in Fig. 2.4. We have chosen the form stated by Barone[28] for the revenue curve.

In our case, the revenue curve follows a path between C and D above the total cost curve. Here, profit is positive. We now look for the point at which profit is highest. As the fundamental proposition implies, this point is characterised by being the point where the two tangents to the revenue curve and the cost curve are parallel to each other for the same velocity of production. In order to find this point, we fill the whole area with curves that are parallel to the revenue curve[29]. The tangents at points on the family of curves that have common abscissae are parallel. One of these curves is tangential to the total cost curve, that is, it has a common tangent with it. Contact point S is the required point. Its abscissa s represents the most favourable velocity of production.

$\overline{TO}$ is the net profit[30] and

$\overline{TA}$ is the gross profit.

$\overline{TO}$ can be negative if T lies above O.

$\overline{TA}$ is always positive, that is, T always lies below A.

---

[27] If $G(x)$ is assumed to be fixed as a continuous function of x (see § 6).

[28] Barone-Stähle: 175, Fig. 48.

[29] These curves represent a family of curves with the differential equation

$$dy = E'(x) \,.\, dx.$$

[30] Note that $\overline{TO}$ is the negative value of $\overline{OT}$.

Fig. 2.4 The marginal costs and marginal returns of the profit-making principle

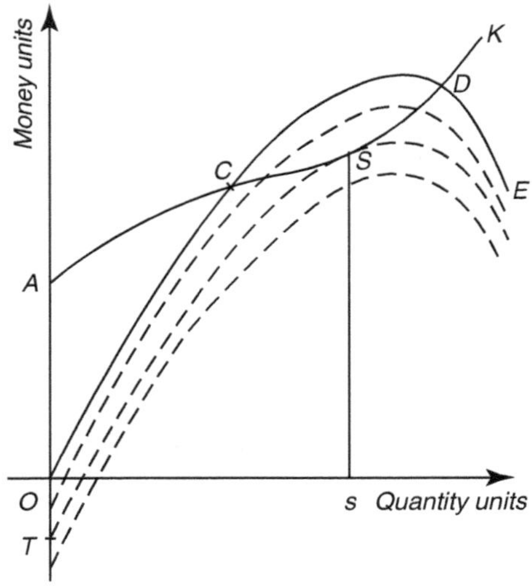

The remarks we have made to prove our fundamental proposition lead straight into another proposition:

(XI). For velocities of production which are lower than the most favourable velocities (however close to *s* they may be), the marginal return is higher than marginal costs. But for velocities of production which are higher than the most favourable velocities (even if they also hardly differ from *s*), the marginal return is lower than marginal costs.

On a graph (see Fig. 2.4), this means that to the left of point *S*, the revenue curve increases more quickly than the total cost curve, whereas it increases less quickly to the right of this.

From this, it follows that it is possible to achieve a particular velocity of production only when there are velocities of production for which marginal returns are higher than marginal costs, and higher velocities of production for which marginal returns are lower than marginal costs. These conditions are still not sufficient however in so far as there must always be velocities of production whose variable costs are lower than an enterprise's returns, so that any production is possible. This is because the enterprise can never incur a loss which is greater than the amount of its constant costs.

In order to consider the necessity of the conditions presented above, we shall investigate the consequences of these conditions not being fulfilled. There are two possible cases:

(a) Where marginal costs are lower than the marginal return for any velocity of production. This would mean that the derivative for profit according to the velocity of production—we can describe this derivative as the marginal profit, by analogy with marginal costs and marginal return—would always be positive for sufficiently high velocities of production. However, this would further mean that the profit for these higher velocities of production would increase monotonically. Then in order to obtain the greatest profit, the enterprise would have to increase its velocity of production ad infinitum but without achieving its objective. This situation in which the enterprise would have to produce "infinitely" more is clearly unthinkable. It would mean abolishing the principle of scarcity. An important proposition emerges from this:

(XII). It is impossible to have a situation in which the marginal costs for all velocities of production exceeding a certain velocity of production are lower than the marginal return.

(b) The second case is where marginal costs for all velocities of production are higher than the marginal return.

Here, returns would always be lower than variable costs. The enterprise would incur the smallest loss if it were to lie idle. We hence obtain the following proposition:

(XIII). Where marginal costs for all velocities of production are higher than the marginal returns, no production can occur at all.

We can hence see that the conditions stated above really must be fulfilled so that some production may occur when the profit-making principle is applied.

In general, we can summarise the results of our study in the following proposition:

(XIV). For any production at all to occur there must be a velocity of production other than zero which maximises profit and which represents the upper limit of the profit function.

If the entrepreneur deviates downwards from the velocity of production at which marginal returns and marginal costs are equal, he finds himself in a situation where the increase in revenue is greater than the increase in costs. This means he makes a profit (Translator's note: in the original German, the author uses the verb "entgeht" which translates as "fails to make [a profit]". From the context of this and the following two sentences, we conclude that the author means the opposite, "makes [a profit]", because he wants to draw a contrast here with the situation where the entrepreneur "entsteht" or incurs [a loss]). If he deviates upwards from the given velocity of production, he arrives at a situation where the increase in costs is greater than the increase in revenue. As a result, he incurs a loss.

4. We must establish another important consequence of the fundamental proposition. The definition of s is revealed using this equation:

$$E'(s) = K'(s).$$

This equation is however completely independent of the amount of constant costs. These can have any random value without changing s. Since the constant costs for all velocities of production have the same value, the total cost gradient is identical to the variable cost gradient. We obtain the following proposition:

(XV). Constant costs are irrelevant when determining the most favourable velocity of production s (they only influence the size of the profit for velocity of production s).

Using equation $K'(x) = K'_{\text{II}}(x)$, our maximisation problem is identical to the definition of the greatest difference between returns and variable costs. It is hence the same whether we ask which velocity of production maximises net profit or which velocity maximises gross profit. We therefore want gross profit to mean the difference between returns and variable costs. All constant costs are hence included in gross profit[31].

It also follows from this that if we arbitrarily make allocations of gross profit to net profit and constant costs, this changes nothing when determining the most favourable velocity of production. The minimum position also remains unaffected by making allocations in such an arbitrary way. It is only the optimum position that also depends on constant costs.

II.

1. We now look at the case of free competition. We have defined this[32] as being a market situation in which the price can be considered as being independent of supply, that is, independent of the enterprise's velocity of production. Here, the return is then the product of the arbitrarily variable velocity of production and the constant price. It is a linear function of the velocity of production and is proportional to this. The proportionality factor (Translator's note: "proportionality factor" see Peacock: 49) is the price.

The marginal return is simply the market price[33]. On the basis of the fundamental proposition of the profit-making proposition for competitive supply, the following proposition then results:

(XVI). In the competitive economy, the most favourable velocity of production is the one where marginal costs equal price[34].

We would like to illustrate this situation using a graph, particularly as constructing the most favourable velocity of production for free competition is very easy. Since the price is constant, the revenue curve is represented as a straight line through the origin using directional tangent P. The most favourable point S on the total cost curve is determined by drawing a tangent to the total cost curve which

---

[31] See Marshall, A. ibid.: Book V.

[32] See Chap. 1, § 2, IV, 1.

[33] It is $E(x) = x.P$ and therefore $E'(x) = P$. (Translator's note: for the term "market price", see Cassell, A. (1918): 180.)

[34] This proposition expresses an old established truth, see e.g. Cournot, A. (1924): 48 (Chap. 8, 2nd equation), that simply expresses something as a formula here, already set out by Ricardo, D. See also, Amoroso, L. ibid.: 9, *Ricardosche Gleichgewichtsformel* (Ricardo Equilibrium Solution).

is parallel to E. Its abscissa is also revealed by the intersection point of the marginal cost curve with the price curve, which is simply parallel to the abscissa axis at distance P. Proposition XI is modified for free competition as follows. Since the marginal return in our example is the gradient of the tangent to the total cost curve at point S, we can say that the total cost curve rises less steeply than the tangent to the left of point S, whereas to the right of it, it rises more steeply. However, this is only possible if the total cost curve in the vicinity of point S is convex downwards, in other words, if it is now subject to the law of diminishing returns. S is hence always greater than b. We would still like to investigate the consequences of the condition of diminishing returns not being fulfilled, that is, if production is subject to the law of either increasing or constant returns.

(a) There are two possibilities where production is subject to the law of increasing returns:

α) Either marginal costs are lower than the price for any velocity of production, in which case the conditions for proposition XII exist. A situation such as this is therefore impossible.

β) Or else marginal costs are higher than the price for all velocities of production. Then the conditions for proposition XIII exist, that is, production cannot occur in this case at all.

Once we have assumed a competitive profit-making economy, we can no longer assume that production is subject to the law of increasing returns or vice versa.

(b) A similar situation exists with regard to the law of constant returns. Here, marginal costs are either higher or lower than the price. In the first case, the conditions for proposition XII similarly exist and in the second case, we have the conditions for proposition XIII. And because it is always possible to have a price which exceeds (constant) marginal costs, we can also say here that the conditions of the "competitive profit-making economy" are irreconcilable with the "law of constant returns".

We hence obtain the following important proposition:

(XVII). A competitive profit-making economy is incompatible with production that is subject to the law of either increasing or constant returns.

This proposition only applies in formal terms, irrespective of the length of the Marshall period of time. Its importance is increased by this fact because the law of increasing returns will probably only apply to an individual enterprise in specific circumstances, on condition that the indirect factors of production remain unchanged[35]. It becomes increasingly important however if we extend our view over a long Marshall period of time and examine production supposing that all factors of production are variable, and if we therefore compare all possible input levels against each other.

We can augment the examination carried out a moment ago by another proposition:

[35] See Bücher, K. (1921b): 95 and 98 (example of the law of increasing returns).

(XVIII). If an enterprise is to function at every price level in the competitive profit-making economy, its marginal costs must increase beyond all limits with an increasing velocity of production.

If this were not the case, there would be an upper limit for marginal costs and so where a price exceeded this upper limit, the impossible situation described in proposition XII would exist.

We gain the following insight from these propositions. A competitive profit-making economy can have production opportunities which are both theoretically possible and which are subject to the law of either increasing or constant returns. These opportunities are therefore only theoretically possible because the corresponding marginal cost functions for all velocities of production exceed the price. However, if the price increases, circumstances may occur in which the production opportunities which are theoretically possible can no longer remain so. For these opportunities, the competitive organisational structure for social production must then deviate towards another organisational structure. The same applies mutatis mutandis for production opportunities which are indeed subject to the law of diminishing returns in which marginal costs however have an upper limit exceeded by the price.

3. We can make another assertion about the position of the most favourable velocity of production. We know that the price must be higher than average variable costs in the minimum position if production is to be possible. Therefore, the marginal costs of the most favourable velocity of production are also higher than the marginal costs of the minimum velocity of production. Since both velocities of production correspond to the increasing branch of the marginal cost function, it hence follows that the most favourable velocity of production must always be higher than the minimum velocity if anything is to be produced at all. We hence obtain the following proposition:

(XIX). If a profit-making enterprise is to produce anything in the competitive economy, the price must then be higher than average variable costs in the minimum position. The most favourable velocity of production that is achieved is then higher than the minimum. This represents the lower limit (Translator's note: "lower limit" see Cassel: 63) of all the possible most favourable velocities of production.

This proposition can also be formulated in a shorter form as follows:

(XIXa). The most favourable velocity of production displays progressive variable costs in the competitive economy.

This fact can also be seen from Fig. 2.5. Here $\overline{TO}$ represents net profit[36], and $\overline{TA}$ represents gross profit. $\overline{TO}$ can also be negative. This is then the case if T lies above the origin. $\overline{TA}$ is always positive. As a result, T always lies below A. However, we know that tangents to the total cost curve which meet the ordinate axis below point A, correspond to points that lie to the right of the minimum position Q.

[36] See footnote to explain net profit (see German ed. pp. 39, footnote 1).

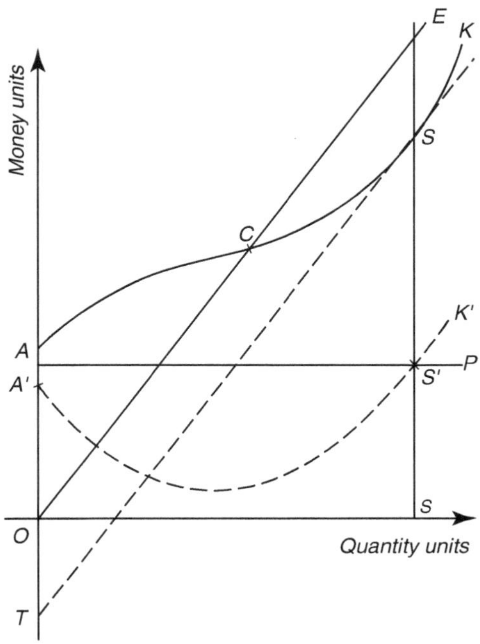

Fig. 2.5 The most favourable velocity of production

Since variable costs become more similar to total costs the longer the Marshall period of time is, we can therefore see that the lower limit of the most favourable velocities of production always comes closer to the optimum position in the long run (the dependency of the contents of our formal propositions on the Marshall period of time must be repeatedly noted and emphasised).

The situation of the most favourable velocity of production is different within the limits stated here. It is dependent on price. Where the price is lower than average costs in the optimum position, the most favourable velocity of production hence lies between the minimum position and the optimum position. Net profit is negative. The enterprise now incurs a loss. One part of constant costs has still not been met. Where the price is higher than average costs in the minimum position, the most favourable velocity of production lies on the other side of the optimum position. Net profit is positive. The enterprise has a net profit which exceeds constant costs. The most favourable velocity of production then coincides, but only with the optimum position if the price is equal to the minimum of average costs. Net profit is now zero. Gross profit is equal to constant costs.

This observation shows us that we can either look at the enterprise's most favourable velocity of production or the enterprise's actual supply in the unit time as a function of price if a competitive economy exists. A certain favourable velocity of production s corresponds to every price and this velocity is calculated from the equation $K'(s) = P$. If this equation has several roots which also satisfy the second maximum condition[37], then the root that is always selected is the one which results in the greatest profit. A definite covariance matrix of the most favourable velocity of production s and price P hence exists. Where the total cost curve is regular, this function is identical to the inverse function $K'(x)$ for all $x > q$.

As a result, we have obtained a new function:

$$s = s(P).$$

This function is the enterprise's supply function. It indicates the velocity of production the enterprise will achieve for a given price and what it will bring to the market. As the inverse function to the marginal cost function, it is monotonically increasing for $x > q$. On the basis of proposition XVII, supply functions that are either monotonically decreasing or constant are incompatible with the competitive profit-making economy. Proposition XVIII must also be seen in this context.

III.

1. In part, we obtain entirely different results if we assume that the enterprise enjoys a monopoly position in its market. Now the price of a good produced and supplied by the enterprise is a monotonically decreasing function of the velocity of production[38].

Here, price and marginal return are different. The geometric description of the situation is rather complex so we must make a preliminary examination in Fig. 2.6.

$C\widehat{C_1}P$ is the demand curve, that is, the price curve. The return for an arbitrary velocity of production $\overline{OD}$ is $\left(\overline{OD}\,.\,\overline{DC}\right)$, hence the surface area of rectangle $ODCF$. An arbitrary alternative (greater) velocity of production $\overline{OD_1}$ has the return $OD_1\,C_1\,F_1$. The increase in revenue is

$$OD_1\,C_1\,F_1 - ODCF = DD_1\,C_1\,H - F_1HCF.$$

We now construct rectangle $GG_1\,C_1\,H$ which has the same surface area as rectangle F_1HCF. We achieve this by pulling F_1G parallel to CC_1. Then the two triangles HC_1C and HF_1G are similar because they have the same angles. Consequently, the following proportion applies: $\overline{HC_1} : \overline{HC} = \overline{HF_1} : \overline{HG}$ or this product equation:

$$\overline{HC}\,.\,\overline{HF_1} = \overline{HC_1}\,.\,\overline{HG}.$$

[37] See proposition (XIX).

[38] See Chap. 1, § 2, IV, 2.

Fig. 2.6 A geometric description of an enterprise's monopoly position

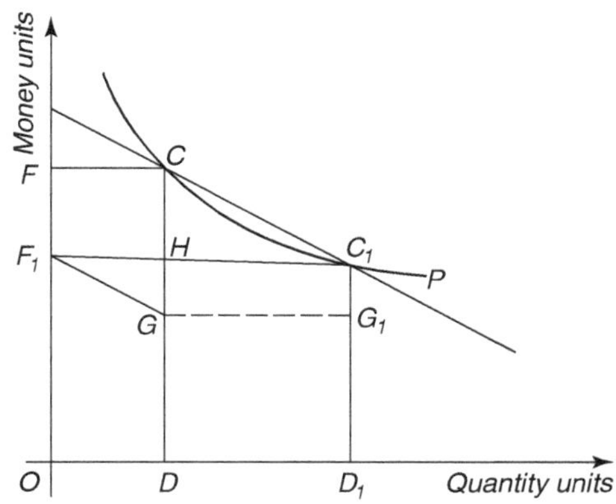

The increase in revenue is therefore rectangle DD_1G_1G if velocity of production $\overline{OD}$ increases by $\overline{DD_1}$.

The size of this increase in revenue is the rectangle's surface area DD_1G_1G divided by the increase in the velocity of production, thus by $\overline{DD_1}$. It is however $\frac{\overline{DD_1}.\overline{DG}}{\overline{DD_1}} = \overline{DG}$. We obtain the marginal return of velocity of production $\overline{OD}$ by allowing $\overline{DD_1}$ to converge towards zero, that is, D_1 moves towards D. Then the secant $\overline{CC_1}$ becomes the tangent of the price curve at point C and F_1 coincides with F. The construction of the marginal return is then revealed accordingly. Figure 2.7 shows how this is constructed.

FG is parallel to CT and DG is the marginal return of velocity of production OD. By constructing this for each point on the abscissa axis, we obtain the marginal revenue curve that corresponds to price curve $\overset{\frown}{CP}$. The marginal revenue curve follows a path completely below the price curve because the price curve is monotonically decreasing. The same hence also applies to its tangents. Therefore, each point G lies below the corresponding point C.

If we describe the tangent of the acute angle between the tangent to the price curve and the abscissa axis as the price differential, we obtain a particular term for variable $\overline{GC}$. It is

$$\overline{OD} = \overline{FC}$$
$$\sphericalangle CTO = \sphericalangle GFC$$
$$\text{tg}\,(\sphericalangle GFC) = \frac{\overline{GC}}{\overline{FC}}$$
$$\overline{GC} = \overline{FC}\,.\,\text{tg}\,(\sphericalangle GFC) = \overline{OD}\,.\,\text{tg}\,(\sphericalangle CTO).$$

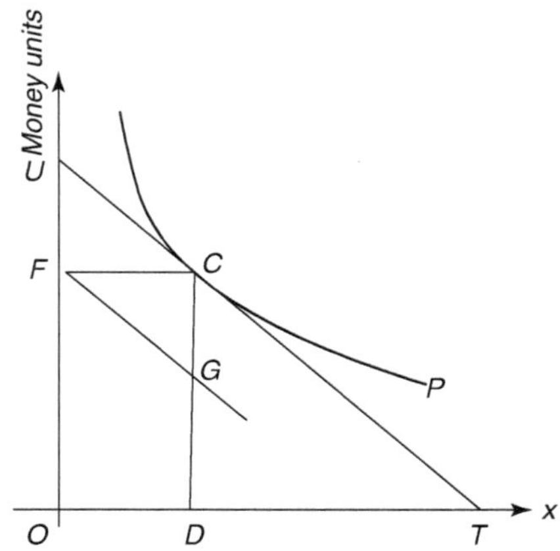

Fig. 2.7 The construction of the marginal return

The marginal revenue curve therefore passes below the price curve by the amount of the product of the velocity of production and the price differential. We can still obtain another term for $\overline{GC}$. As usual[39], we refer to the quotients $\frac{\overline{TC}}{\overline{CU}}$ as the elasticity of demand at point C. Now the triangles TCD and FGC are similar because their angles are equal. Therefore, the following formula applies using proportion $\overline{GC} : \overline{GF} = \overline{DC} : \overline{TC}$ and $\overline{GF} = \overline{CU}$:

$$\overline{GC} = \overline{DC} : \frac{\overline{TC}}{\overline{CU}}.$$

This means to say that the marginal revenue curve is shifted downwards towards the price curve by the amount of the quotient of the price and the elasticity of demand[40].

These preliminary remarks, in conjunction with the fundamental proposition of the profit-making principle, immediately lead to the following three propositions.

(XX). The most favourable velocity of production of a monopolised profit-making enterprise is the one where marginal costs equal price, minus the product of the velocity of production and the price differential.

(XXa). The difference between the price and the marginal costs of the most favourable velocity of production of a monopolised profit-making enterprise

[39] See Dalton H., *The Inequality of Incomes*: 192 et seq.

[40] Constructing the marginal revenue curve is particularly easy if the price curve is linear—the marginal revenue curve is also linear. Then we only need to construct one point on the marginal revenue curve. This point's connecting line (Translator's note: see Peacock, A.: 310) with the intersection point of the price curve (which is a straight line here) and the ordinate axis, is the required marginal revenue curve.

equals the price of this velocity of production divided by the elasticity of demand[41].

(XXI). In the case of a monopoly orientated towards profit-making production, the price achieved is always higher than the marginal costs of the most favourable velocity of production.

We would like to add a further comment about the proposition just formulated, namely that the greater the amount by which the price exceeds marginal costs, the lower the elasticity of demand is. By contrast, where the elasticity of demand is very high, the price is almost equal to marginal costs. Here, we have an approximation of the assumptions of free competition. We can also in fact study price as a function of the velocity of production in the case of free competition and do this by assigning the same price to every velocity of production. The graph of this function appears as a line parallel to the abscissa axis. A demand function with very high elasticity runs on a parallel that hardly differs from this. In this respect, the case for free competition as we have defined it, can be seen as a borderline case of monopoly if the elasticity of demand increases beyond all limits. With regard to the individual enterprise, actual free competition does not represent this borderline case (where we can formally describe elasticity to be infinitely high), but instead represents the monopoly case where the elasticity of demand is very high. The observation may be made for free competition in our sense without this leading to any major mistakes. We must always keep in mind that this concerns a borderline case which only approximately describes reality.

2. We have established above that competitively organised, profit-making production is not always accomplished because the most favourable velocity of production, which maximises profit and hence determines the production level to be achieved, does not always exist under competitive conditions. We can now ask ourselves whether a most favourable velocity of production exists in the case of a monopoly and whether production is hence always completely determined by the prerequisite regulatory principles in this case. The observation below shows us that this question can be answered in the affirmative.

In order to be able to successfully make our observation, we must determine a particular feature of demand as follows. In a fixed market economy, an upper limit always exists for the total amount spent on a particular type of good in the unit time. This assertion is indeed very convincing and is proved by the scarcity principle and the theory of marginal utility (Translator's note: "marginal utility" see Scandizzo: 50). We will not present the proof because it lies outside the scope of this book. We can make this assertion as a requirement for all the demand functions we meet, where an appropriate investigation would show that other demand functions cannot also exist.

[41] See also: Amoroso, L. ibid.: 10 which presents this proposition in a formula as follows:

$$p - m = \frac{p}{\varepsilon}$$

where p is the price (Italian: "prezzo"), m are the marginal costs (costo marginale) and ε is the elasticity of demand.

From this, it follows that profit must also have an upper limit because the return has an upper limit. There are certainly velocities of production where the difference in economic terms between profit and the upper profit limit is so small it can be ignored (e.g. 0.0001 German Pfennig (Translator's note: old German currency: a penny)). Each of these velocities of production is a "most favourable" velocity, assuming it is worth the enterprise producing anything at all.

We therefore arrive at this important proposition:

(XXII). Monopolistically organised production orientated towards profit-making always succeeds.

This proposition represents a fundamental difference compared with production in a competitive economy (see proposition XVIII). Provided the profit-making principle applies, it follows from this operational guarantee that some branches of production have the choice between a competitive and a monopolistic structure, whereas others have to rely only on a monopolistic structure. A competitively organised branch of production must hence change into a monopoly as soon as the production conditions change accordingly[42]. We are also able to indicate the way in which any such organisational change could occur. Where an enterprise that is largely subject to the law of increasing returns e.g. for all velocities of production that are generally possible to satisfy demand, enters a branch of production, it drives all other enterprises out of the market by expanding its production and achieves a state of monopoly for itself. Another structure would, for instance, be one where either all or most enterprises of a branch of production are subject to the law of increasing returns to a greater extent by developing their productive powers and, so that they can be kept informed about the general situation, they form a cartel.

IV.

We still need to concentrate on modified rivalry[43], that is, the case in which price is a constant and the sales quantity depends on sales costs, whereas conversely, production costs are determined by the sales quantity. The problem that exists here is again that of determining the most favourable velocity of production. We are looking for a velocity of production whose production costs, increased by the sales costs required to sell the entire quantity of products produced in the unit time, are exceeded by the maximum amount of return, which is simply the greatest possible profit.

[42] It can be shown that an intermediary stage, that is, free competition involving a few enterprises, is not possible without certain additional assumptions. (Then see Edgeworth, F.Y. and Pareto, V.; by contrast, see Cournot, A. (1924); Schneider, E. (1930); see Amoroso, L. ibid.: 13 et seq.) This problem has been dealt with well and in great detail by Sting, K. (1931): 761 et seq. The descriptions he chooses may not be very satisfactory. Too little importance is attached to "hyperpolitical pricing" (Translator's note: "Preisbildung", see Cassel, G. "pricing": 102). Sting may be too optimistic about the supplier's "polypolitical" inclination. See also: Crosara, A. (1930): 25 et seq.

[43] See Chap. 1, § 2, IV, 3.

Now total costs have the following structure, as we have shown for the definition of modified rivalry: $K(x) + C(x) = \mathfrak{K}(x)$. By treating $\mathfrak{K}(x)$ in exactly the same way as K (x), we arrive at this proposition:

(XXIII). The modified competitive economy is subject to exactly the same laws as the pure competitive economy if we regard the sum of production costs and sales costs to be the relevant enterprise's total costs.

Based on our assumptions, since there should be a link between the quantity sold and the quantity produced, A (x) then formally has exactly the same meaning as $K(x)$ for ordinary free competition. As our propositions are all formal, this leads to the assertion put forward a moment ago.

The case of modified rivalry should be particularly emphasised because it is often shown in practice that firms fail to reach the most favourable velocity of production because of a lack of sales. Here, sales costs must be included if the "marginal costs equal price" law is to still be fulfilled.

§ 5. The Enterprise's Supply According to the Principle of the Satisfaction of Needs and Wants

We shall now replace this profit-making principle with the principle of the satisfaction of needs and wants in order to examine the consequences which arise from the composition of this principle with regard to the other remaining unchanged assumptions. We would therefore like to trace the effects of this principle in the competitive economy, the monopolistically organised economy and the modified competitive economy in turn so as to then turn to the special case where the price is initially uncertain and only the quantity ordered is certain.

I.

It follows from the definition of the satisfaction of needs and wants that production, when orientated according to this principle, is usually uncertain in the case of rivalry and also in cases where the price is fixed and the saleable quantity is arbitrary. We already know that a price then covers the total costs of a velocity of production if it is either equal to the relevant average costs or exceeds them. From this it emerges that no production can occur at all where there is no velocity of production with average costs which are either lower than the price or equal to it, whereas every velocity of production with average costs that are either equal to the price or lower than the price is achievable according to this principle. As formulated above, from this principle we cannot deduce what should now really be obtained for these velocities of production. Production is only unambiguously determined via this principle in special exceptional cases where a single velocity of production does not have higher average costs than the price. This velocity of production can clearly only be the optimum velocity. Here price and average costs would be exactly equal to each other. The enterprise would hence achieve its optimum position on the basis of the principle of the satisfaction of needs and wants. Otherwise, we would only be able to clearly determine production using an additional principle. For example, we could determine that the enterprise should always achieve the optimum position without taking the pricing level into consideration, provided that any production at all can occur. Alternatively, we could establish that the greatest possible quantity should be supplied at the price concerned. Each of these two secondary principles would produce a definite determination of production in many circumstances and so enable a definite socio-economic equilibrium. However, they would break down in the case of cost degression. Moreover, the second principle would also break down if the average costs of a certain velocity of production perhaps increased but continually remained below the price. In these circumstances, the competitive organisation of production would not be possible.

If we do not assume these subsidiary principles, the competitive organisation of production then becomes possible in the sense that there is no tendency towards an excessive expansion of production which would end competition. However, price is therefore no longer suitable for balancing demand and supply because it cannot clearly determine supply, but since the possibility of looking at free competition in the form of "constant price, arbitrary supply" is based on the assumption of a general equilibrium, we can establish the following proposition:

(XXIV). The principle of the satisfaction of needs and wants, excluding other subsidiary principles, is only consistent with the assumption of free competition in special circumstances.

In addition the two subsidiary principles we mentioned above do not always result in an acceptance of the principle of the satisfaction of needs and wants and free competition. In particular, this compatibility is missing in all circumstances where the assumption of the profit-making principle is missing.

II.

It is different for the case of monopolistically organised production. The following proposition applies:

(XXV). The principle of the satisfaction of needs and wants in conjunction with the subsidiary principle, which states that as much should be produced as possible, is always sufficient to determine production and establish the economic equilibrium in a monopolistically organised economy.

This proposition would only fail to apply if the revenue function alone had an upper limit that was not a function value and would be closer to this function with an increasing velocity of production, although the total cost function would also have an upper limit and this would not be greater than the upper limit for the return[44].

However, we can describe this as impossible because then there might be a velocity of production above which total costs might differ only by a negligible amount in terms of their upper limit[45]. We could then say that total costs are constant above this. And yet, well-established practices teach us that a situation like this cannot happen.

In all other cases, there is one velocity of production at which average costs and price are equal to each other and above which the price is lower than average costs. This velocity of production is achieved on the basis of the principle of the satisfaction of needs and wants and the subsidiary principle mentioned above.

III.

In contrast to the results we have obtained by assuming the profit-making principle, which show us that there is no significant difference between the situations under free competition and those under modified competition, a special situation arises for modified competition in the case of the principle of the satisfaction of needs and wants. In contrast to free competition, here the sales volume is firstly fixed. It can be expanded via expenditure on sales costs. However, for the principle of the satisfaction of needs and wants, there is no reason to especially promote sales. As a result, the velocity of production is now regarded as a given. It is equal to the sales quantity which occurs when the sales costs are zero. This velocity of production is achieved if the relevant average costs are not higher than the price.

On this occasion, we shall give a more detailed interpretation of the principle of the satisfaction of needs and wants for the two cases where:

1. No price can be achieved which would be equal to any level of average costs
2. The quantity ordered cannot be supplied at the price offered.

The first case is important for all market situations while the second is only important for modified competition.

1. In a situation where costs cannot be covered and a loss is incurred, striving to meet costs must change to become striving to incur the smallest loss possible.

---

[44] Expressed as a formula, this must be:

$$\lim_{x\to\infty} E(x) > E(x) \text{ for all } x$$
$$\text{and } \lim_{x\to\infty} K(x) \leqq \lim_{x\to\infty} E(x).$$

[45] See § 4, III.

This means that whenever no velocity of production at all can be achieved according to the principle of the satisfaction of needs and wants, the profit-making principle must intervene to replace the principle of the satisfaction of needs and wants.

2. When the quantity ordered cannot be delivered at the price offered, striving to supply the quantity ordered must have the effect from the demand side of driving the enterprise (as much as possible), in the direction of supplying a quantity which differs as little as possible from the quantity ordered and whose costs are covered by the price offered. If the velocity of production cannot be achieved at the price offered, there are however lower velocities of production which could be achieved based on the observations in section I. The highest of these is hence achieved from the demand side.

It might incidentally also be possible that the enterprise would try to achieve a greater velocity of production in similar circumstances (e.g. under the assumptions of subsidiary principles) via expenditure on sales costs. However we do not need to go into this exception in any detail.

IV.

A market situation which has a special affinity with the principle of the satisfaction of needs and wants, that is, a situation which we have not described yet because it is inconceivable in the case of the profit-making principle, occurs in the following way: a particular quantity is demanded and the price is initially uncertain. From the principle of the satisfaction of needs and wants, it emerges that the price of this quantity is equal to its average costs because it is only then that the cheapest possible supply is met by the production quantity ordered. Now the principle of the satisfaction of needs and wants appears in sharp clarity without the need for subsidiary principles. We will see later that the principle of the satisfaction of needs and wants is perfectly applicable in this sense under certain circumstances based on the profit-making principle[46].

References

Barone, E. (1927). *Grundzüge der theoretischen Nationalökonomie* (published in English as: *Fundamental Principles of the Theory of National Economics*), Bonn.

Barone-Stähle, *Grundzüge der Theoretische Nationalökonomie* (published in English as: *Principles of Theoretical Economics*): 175, Fig. 48.

[46] See Theory of the Transfer Price, Chap. 3, § 4.

Brinkmann, Th. (1935). *Economics of the farm business, with introduction and notes by Elizabeth Tucker Benedict, Heinrich Hermann Stippler and Murray Reed Benedict*. Berkeley, CA: University of California Press: 7.

Bücher, K. (1921a). "Gesetz der Massenproduktion" (Law of Mass Production) in *Die Entstehung der Volkswirtschaft* (The rise of the National Economy), 2nd collection. Tübingen, 92.

Bücher, K. (1921b). *Das Gesetz der Massenproduktion* (The Law of Mass Production) in *Die Entstehung der Volkswirtschaft* (Industrial Evolution), 2nd edn. Tübingen, 95 and 98.

Cassel, G. (1918). *Theoretische Sozialökonomie* (published in English as: *The Theory of Social Economy*), 4th edn: 252 et seq.

Cournot, A. (1924). *Untersuchungen über die mathematischen Grundlagen der Theorie des Reichtums* (published in English as: *Researches into the Mathematical Principles of the Theory of Wealth*). German ed., Jena.

Crosara, A. (1930). "Della identità dei concetti astratti di monopolio..." (Giornale degli Economisti, 25 et seq.).

Jevons, W.S. (1924). *Die Theorie der politischen Ökonomie* (published in English as: *The Theory of Political Economy*), Wantig edition, Jena: 198 et seq.

Kalischer, H.E. (1930). "Der Widerspruch zwischen mathematischer und buchtechnischer Kostenauflösung" (The Contradiction Between Mathematical and Accounting Cost Estimation), *Zeitschrift für handelswirtschaftliche Forschung* (Review for Commercial Research), April 1929 and especially January: 18 et seq.

Pareto, V. (1897). *Cours d'économie politique* (published in English as: *Manual of Political Economy*) II, Lucerne, 102 et seq.

Pigou, A. C. (1927). The laws of diminishing and increasing cost. *Economic Journal, 37*, 188.

Ricardo, D. (1921). *Grundsätze der Volkswirtschaft und Besteuerung* (published in English as: *On the Principles of Political Economy and Taxation*), Wantig edition, Jena, 52 et seq.

Schneider, E. (1930). *Sozialwissenschaft und Sozialpolitik* (Social Science and Social Policy).

Schneider, E. (1931a). Kostenanalyse als Grundlage einer statistischen Ermittlung von Nachfragekurven (Cost analysis as a basis for the statistical determination of demand curves). *Archiv für Sozialwissenschaft und Sozialpolitik, 66*, 585.

Schneider, E. (1931b). "Zur Interpretation von Kostenkurven" (Interpreting Cost Curves). *Archiv für Sozialwissenschaft und Sozialpolitik, 65*, 269.

Stanley, J. W. (1871). *The theory of political economy* (p. 200). New York, London: Macmillan. XVI+267.

Sting, K. (1931). *Die polypolitische Preisbildung* (Polypolitical Determination of Prices), Jahrbücher für Nationalökonomie und Statistik, 761 et seq.

Chapter 3
Costs in Joint Production

The situation we have examined so far, where only one good is produced, in reality plays a role which is not to be underestimated because, like every economic theory, it too only really applies where the conditions are broadly met. For example, if a by-product is produced in addition to the main product and this provides only a small fraction of the revenue, we can then safely apply the theory of "single" supply. We therefore deduct the revenue of the by-product from total costs to perhaps achieve greater precision and regard the difference as being the principal product's total costs.

However, frequently this is not possible. Multiple goods are produced simultaneously and these are more or less equally important. The theory developed so far is no longer adequate and we must therefore use a more general theory, that is, the theory of joint supply.[1]

§ 1. Theory of the Length of Production

I.

The subject of this chapter is more complex than that of the previous chapter. We hence choose the simplest conditions possible and only present those which are fundamental. We shall only examine the case of two goods being produced because this already provides all the methodological ideas which must be used in a general example where n goods are produced. We shall then make the assumption that all functions occurring in the area under consideration are regular and smooth, that is, they are continuous and can be differentiated. There should be no jumping costs. First of all, we use a simple approach. Any arbitrary combination of quantities of

[1] See Fanno, M. (1914). It would be too complicated to compare and contrast this paper because our explanations have completely different basic ideas and structures.

H. von Stackelberg, *Foundations of a Pure Cost Theory*,
DOI 10.1007/978-3-642-34537-1_3,

Fig. 3.1 Cost table for joint production

Quantity of good 2 \ Quantity of good 1	0	1	2	3	4	5
5	61.9	64	71.3	87.2	116.6	164.5
4	58	59.6	65.6	79.4	106.5	151.5
3	55.6	56.9	61.9	74.2	99	141.8
2	54	55.3	59.6	70.6	93.8	135.2
1	52.4	54.4	58.4	68.6	91.1	131.7
0	50	54	58	68	90	130

good 1 and good 2 has particular production costs which must be measurable in some way. In order to provide a clear description, we draw up a table (Fig. 3.1).

The quantities of each good decrease on a suitable scale in the title row or column outside the table (which, by analogy with the coordinate representation, has now been set downwards) so that the distance from the middle of this title row or column to the middle of a row or column defines the row or column heading. The dotted lines correspond to the coordinate axes. Their point of intersection is the origin, that is, the firm's idle point. The costs of the relevant combinations of quantities are inserted into the boxes. So, producing 3 quantity units of good 1 and 2 quantity units of good 2, costs 70.6 money units in our example, etc. Each individual point on this table is a particular combination and therefore has particular total costs.

In our table we have determined a point by giving the distance from the title row to the title column, and the distance from the title column to the title row. We shall now use a different method. We would like to measure each point's distance from the origin, that is, from the point (0, 0) and either describe this distance as the length of the relevant production combination or, as we have already termed it, the length of the relevant product vector. Secondly, we would like to determine the angle between the line which connects the origin with a point on the table and the horizontal and we will describe this angle as the direction of the product vector. It is easy to see that the product vector is clearly determined by giving its "length" and its "direction". From the table we can then immediately read off the quantities of the two goods. Vectors of the same length define points that are equidistant from the origin, that is, the sum of the squares of the quantities produced of goods 1 and 2 is equal for these points. The points lie on a circle around the origin with a radius that constitutes the length. Points in the same direction lie on a line which itself goes through the origin. They are characterised by the fact that the proportion of the two goods (that is, the components of the product vector) is the same. We describe the length using r and the direction using φ. In the following pages we will use the system of coordinates instead of the table. On the abscissa axis we have the quantities of good 1 and on the ordinate axis we have the quantities of good 2.

Fig. 3.2 Costs as a function

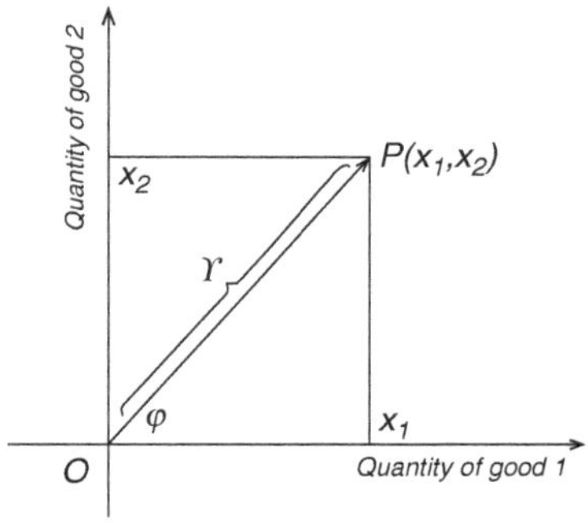

We can imagine total costs (and similarly, other cost functions) as being measured on a third axis and determined by a perpendicular to the plane of a piece of paper. Costs appear to be a function of the quantities of goods 1 and 2 or even a function of the "length" and "direction" of the existing product vector.

Fig. 3.2 illustrates the situation described here. The following relationships apply:

$$r^2 = {x_1}^2 + {x_2}^2; \; \tan \varphi = \frac{x_2}{x_1}.$$

Two variables can each be calculated from these two equations if the other two variables are given.

To be consistent with the wording in normal usage, we describe r and φ as polar coordinates.

Let us look at the case where the ratio in which both goods are produced is fixed, that is, the proportion $x_1 : x_2$. This means that the direction is invariable. φ is constant. Production is only adjusted by changing the "length". In this case, determining the production is exactly the same as if it were a case of single production. The fact that two different goods are produced is not at all important here for the enterprise. Let us imagine that there is a particular quantity of good 1 produced and that this quantity is combined with the quantity of good 2 produced and that they are then aggregated and described as a packet (Translator's note: "packet" see Baumgärtner: 514). We can hence say that the enterprise produces a particular number of "packets" in the unit time. This "packet" appears here as a quantity unit. The number of "packets" produced in the unit time is the velocity of production. Costs and returns only depend on this now because the composition ratio of the "packet" itself remains unchanged.

Nothing has been changed for "single" supply, not even in terms of the market situation which this enterprise faces. Where the price of any good is a constant, the price of the "packet" is obtained by multiplying the quantities of goods contained in the "packet" by the corresponding prices and adding the results. A product vector's return is revealed by multiplying the price of a "packet" by the number of "packets". Where price is dependent on the velocities of production of individual goods, a particular price for each good and also therefore for each "packet" hence corresponds to every product vector. However, as the product vector is none other than a certain number of "packets", price is only dependent on this number, which also corresponds here to the particular velocity of production in the case of single production.

If we again establish that this "packet" is such that the sum of the squares of the quantities of goods contained in it equals 1, then the number of "packets" is always identical to the "length" of the product vector and hence to r (Translator's note: this sentence, and "equals 1" especially, has been much discussed by the translators and now reflects von Stackelberg's probable intended meaning). In the following pages, we shall describe the "packet" itself as a "unit vector" (Translator's note: see Baumgärtner: 514) and denote it using the symbol $\mathfrak{e}$. Since $\mathfrak{e}$ is clearly determined when the direction φ is specified, $\mathfrak{e}$ can hence be considered as being a function of φ. In contrast to the right-angled coordinates (x_1, x_2), we point to the dependence of a variable on the polar coordinates r und φ using square brackets [ ]. Then $\mathfrak{e} = \mathfrak{e}\,[\varphi]$ applies. Each product vector $\mathfrak{x}$ is now represented as the product of its length and the unit vector of its direction[2]:

$$\mathfrak{x} = r\,.\,\mathfrak{e}\,[\varphi].$$

Where the direction is fixed,[3] the unit vector is also fixed. The product vector then only depends on its length.[4] It then follows from this that only the profit, return

[2] This vectorial relationship is clearly visible. From Fig. 3.2 we can see that these equations apply:

$$x_1 = r\,.\,\cos\varphi;\; x_2 = r\,.\,\sin\varphi.$$

It is hence

$$\mathfrak{r}_1 = (x_1, x_2) = (r\,.\,\cos\varphi,\, r\,.\,\sin\varphi) = r\,.\,(\cos\varphi,\, \sin\varphi).$$

On the one hand, since the equation $e_1{}^2 + e_2{}^2 = 1$ applies to vector $\mathfrak{e}$, whose components we describe with e_1, e_2; on the other hand, if we describe the length of $\mathfrak{e}$ using $|\mathfrak{e}|$, we obtain:

$$e_1 = |\mathfrak{e}|\,.\,\cos\varphi;\; e_2 = |\mathfrak{e}|\,.\,\sin\varphi$$

and we hence derive:

$$|\mathfrak{e}|^2 = |\mathfrak{e}|^2\,.\cos^2\varphi + |\mathfrak{e}|^2\,.\sin^2\varphi = 1.$$

Since we only know positive lengths, $|\mathfrak{e}| = 1$ and as a result $\mathfrak{e} = (\cos\varphi, \sin\varphi)$, then ultimately

$$\mathfrak{r} = r\,.\,\mathfrak{e}\,[\varphi].$$

[3] Hence φ = constant.

[4] For an arbitrary function Φ $(\mathfrak{x})$ of product vector $\mathfrak{x}$, we hence derive this relationship:

$$\Phi(\mathfrak{x}) = \Phi(\mathfrak{e}\,.\,r) = \Phi[r].$$

and total costs are functions of the length.[5] Here, length plays exactly the same role as the velocity of production in single production. We therefore obtain the fundamental proposition of joint production (Translator's note: "fundamental proposition of joint production" see Baumgärtner (2001): 514):

(XXVI). With a fixed proportion of the goods produced, all laws of single production apply to joint production where the velocity of production for single production is replaced by the length (the absolute amount) of the product vector.

Single production appears to be a special kind of joint production with a fixed proportion if we let either the second component of the unit vector or even φ equal zero.

Any functions which occur appear to be functions of the point on the line which is specified by unit vector $\mathfrak{e}\ [\varphi] = (\cos\ \varphi,\ \sin\ \varphi)$. Accordingly, a point b, and a minimum and an optimum position correspond to every direction φ, as well as a most favourable velocity of production with a given revenue function. We will refer to these variables in the same way we did in Chapter Two by always indicating the direction in this fashion and placing φ after it in brackets. These variables are simply functions of φ.[6]

Where the price of individual goods is independent of the velocity of production, we should nevertheless note that the price of the unit vector depends on the direction. We refer to the price[7] of the unit vector of a particular direction φ using $P\ [\varphi]$.

These remarks should clarify the fundamental proposition. In principle, the case of joint production and a constant proportion is dealt with by this proposition. It is entirely reduced in the case of "single" supply.[8]

---

[5] We therefore derive:

$$G[r] = E[r] - K[r].$$

[6] We derive:

$$b[\varphi];\ q[\varphi];\ p[\varphi];\ s[\varphi].$$

[7] Then the following applies:

$$P[\varphi] = P_1 . \cos\varphi + P_2 . \sin\varphi.$$

[8] It is indeed very clear here that an allocation of total costs to individual products, or even only variable costs, is completely pointless. The various product combinations appear to be various quantities of one and the same product. Each individual velocity of production costs just as much as the entire product vector if we ask the question, "What must be sacrificed to obtain the relevant velocity of production?" And conversely, each velocity of production costs nothing if we ask the question, "What can we save if we give up producing a good in order to produce other goods in unchanged quantities?"

§ 2. Theory of the Direction of Production

I.

We have just seen how all categories of single production apply in joint production as soon as the "direction" is fixed, that is, the proportion in which the two products are produced is fixed. Let us now assume that the direction is variable. This would mean that our enterprise is not only in a position to adapt to the market conditions in terms of the length of production, that is, in terms of the quantity of products produced with a constant proportion, but it is also in a position to vary this proportion and choose the most favourable proportion from all the possible proportions between x_1 and x_2.

We can then assign a profit function and a revenue function to each direction as well as a total cost function and its derivatives, an average cost function and a function for average variable costs, and these will always depend on the length of the product vector in the relevant direction. Accordingly, we can assign a point b, a minimum position, an optimum position and a most favourable velocity of production to each direction.

All of the points b for all directions and similarly all of the other points that have been mentioned also form a curve. We therefore derive 4 curves: the b curve, the minimum position curve, the optimum position curve[9] and the most favourable velocities of production curve. All of these curves make the radius vector a definite function of the direction. We can hence label the curves using the following terms: $b\ [\varphi]$, $p\ [\varphi]$, $q\ [\varphi]$ and $s\ [\varphi]$. The points distinguished in the one-dimensional case appear to be curves that are distinguished in the two-dimensional case, but with one reservation. The particular "most favourable" velocities of production do not equal each other using $s\ [\varphi]$. We can compare these velocities of production in terms of the profit that can be achieved by them. In reality, the only velocity of production obtained for all velocities of production $s\ [\varphi]$, is the one which produces the greatest profit. Consequently, there is (as a general rule) ultimately just one product vector which is the most favourable.

If we describe its direction using σ, then in reality $s\ [\sigma]$ is the most favourable velocity of production that can be achieved.

Our comments are illustrated in Fig. 3.3.

On the basis of the knowledge gained in Chap. 2, curve $b\ [\varphi]$ lies inside $q\ [\varphi]$ and $q\ [\varphi]$ lies inside $p\ [\varphi]$. By contrast, in the competitive economy, $s\ [\varphi]$ can always follow a path outside the scope of $q\ [\varphi]$.[10] All s which lie inside $p\ [\varphi]$ produce a loss in the competitive economy and all s which lie outside $p\ [\varphi]$ produce a profit.

---

[9] Translator's note: 'Betriebsoptimumkurve' and 'Betriebsminimumkurve' respectively in German, literally 'firm's optimum/minimum curve', but von Stackelberg seems to use this to apply to a firm as well as an enterprise. For clarity, we have opted to use 'optimum/minimum position curve'.

[10] s can only follow a path inside $q\ [\varphi]$ in the appropriate case of the monopoly.

Fig. 3.3 The most favourable product vector

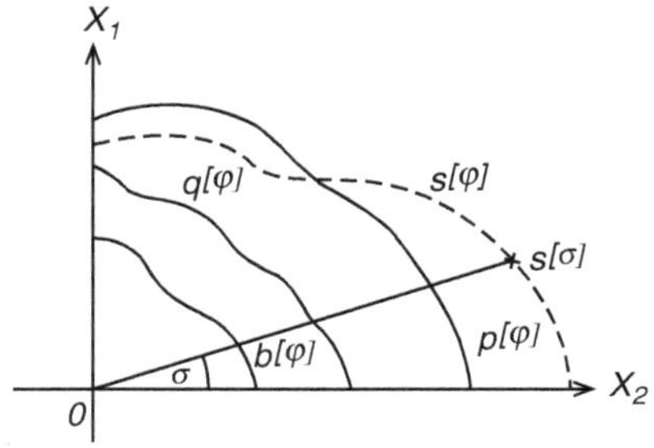

If $s\ [\varphi]$ follows a path partly inside and partly outside $p\ [\varphi]$ as in the figure above, it is hence obvious that the most favourable point $s\ [\sigma]$ can only lie in the part of $s\ [\varphi]$ that follows a path outside $p\ [\varphi]$.

Where free competition exists, we are able to assign a most favourable product vector $s\ [\sigma]$ to every price vector. We hence arrive at the enterprise's supply as a function of the price vector.

The following propositions apply to joint production:

(XXVII). The total cost function of an enterprise with joint production is shaped in such a way that it fulfils the laws of single production in every direction as set out in the fundamental proposition[11] of joint production.

(XXVIII). The curves that can be distinguished in the joint production of two goods correspond to the points that can be distinguished in single production (in general terms, that is $(n-1)$ dimensional diversities with the joint production of n goods). Moreover, the following proposition for joint production leads to:

(XXIX). One most favourable direction of production (Translator's note: "direction of production" see Cassel (1967): 65) correlates to each length of production and as a result, a curve for the most favourable directions is defined. This curve and the curve for the most favourable length of production determine the enterprise's most favourable production vector (Translator's note: "production vector" see Baumgärtner: 516), that is, its supply, using its intersection point.[12]

The definitive proof for this proposition will follow later in proposition XXXI.

We can see that "direction" is the new element being examined for joint production. This also applies to cases where more than two goods are produced jointly. The single production categories are uncertain in terms of the direction. Neither do they allow a logical supplementary definition which could have been clearly determined. Only the most favourable product vector is defined as being the most favourable, not just with regard to length, but direction too.

[11] Proposition (XXVI).

[12] In terms of the principle of the satisfaction of needs and wants, we can now say that an additional subsidiary principle also becomes necessary in the monopoly economy and that of all the highest possible velocities of production (each direction has such a velocity of production), the velocity which should be obtained is the one which has the lowest average costs. Apart from the length, the direction must now also be determined somehow.

The task which we had set ourselves is solved by these explanations of joint production and also for the general case. We can however provide yet another (insignificant) supplement to the fundamental proposition (assuming that all functions which occur are regular) and also consequently for proposition XXVIII. It applies in particular for the case where the direction is not constant, but is also a clearly continuous function of length, as well as the fundamental proposition. In this case, the economic functions (profit, returns and costs) are not defined along a line, but along a curve instead. The length of the respective product vector is maintained as a velocity of production.

The following is the fundamental conclusion for describing joint production so far: if we use a suitable definition for the terms "length" and "direction" of the product vector, then all assertions regarding single production can also be transferred to joint production. Joint production does not appear to be organised like single production, but instead is given precedence over it. Single production is a special kind of joint production. And indeed, this is decisive. The theory of single production is completely contained in the theory of joint production, except that the theory of joint production additionally sets out elements which are not contained in the theory of single production. These elements are closely associated with the concept of the "direction" of the product vector. We can therefore formulate the following proposition:

(XXX). The theory of the velocity of production applies to all production. The theory of the direction of production also applies to joint production.

II.

In the case of joint production, returns and total costs are dependent on two goods. Since we can present velocities of production as functions of the length and direction of the product vector, we can also say that total costs and returns are functions of the length and direction of the product vector.

In the following pages, let us imagine that the length of the product vector is fixed. Then the variables in question are only dependent on the direction of production. In this example we can regard total costs and returns as functions of the direction of production φ. Here, it is now of crucial importance that angle φ can only vary between 0° and 90° and that the functions in question are therefore only defined within a restricted area. If we measure angle φ in radians, we can say that the functions mentioned above are only defined in the interval $\left[0, \frac{\pi}{2}\right]$.

The same also applies to all functions which can be composed or derived from the revenue function and the total cost function and this does not contradict the principle of continuity in any way. The following important proposition results from this fact and also from the assumption made above that all functions which occur should be regular and smooth:

(XXXI). The revenue function and the total cost function, as well as their compositions and derivatives, always have a minimum and a maximum position in the scope of the variable φ.

This hence also applies to the profit function. As a result, there is always a most favourable direction of production regardless of how either the revenue function or the cost function are otherwise obtained (see proposition XXX). In terms of the direction of production, unlike the velocity of production, enterprises do not need to show any particular features of their total cost function to be able to operate.

III.

Again, we would like to remind the reader of the importance of the polar coordinates we used before and take this opportunity to make an important addition to the theory of the direction of production.

The polar coordinates are a special kind of curvilinear coordinate. They consist of a family of lines which pass through the origin and a family of concentric circles that pass around the origin. Each point on the plane is shown to be an intersection point of a circle and a line. Examining the length of the product vector for a fixed direction of production just means examining the functions along a line that interests us. The theory of the direction of production is simply the investigation of these functions along a circle.

We have already mentioned that the same laws which apply to our functions along a line through the origin also apply along a curve which the direction identifies as being a definite continuous function of the length of the product vector. It is therefore easy to see that the examination along a line through the origin is merely a special kind of this general examination. The line in particular identifies direction to be a constant in terms of the length.

Similarly, it is now possible to generalise about observations along a circle. The circle is a curve which defines length as a constant in terms of the direction. The propositions which apply along a circle remain valid if, instead of a circle, we consider another curve which satisfies the requirement that length is defined as a definite continuous function of the direction.

As a result, we can replace the lines with a family of suitable non-intersecting curves and similarly also replace the circles with other suitable curves which must likewise not be intersected.[13] The first possibility is not relevant to us, however the second is. In the following section, we will present a general definition of the most

[13] Each of these families of curves must fulfil the requirement that its curves do not intersect each other. Otherwise the requirement for the invertible definite covariance matrix of points and their coordinates would not be fulfilled.

favourable velocity of production according to the profit-making principle for joint production. Here, we will look at a curvilinear system of coordinates, consisting of lines through the origin as well as curves which define the length as a definite continuous function of direction.[14] As a result, we will successfully obtain the most favourable velocity of production for joint production from the results of single production and also from proposition XXXI of the theory of the direction of production, using a geometric construction that is already familiar in theoretical economics.

IV.

1. A combination of the two velocities of production x_1, x_2 is represented using a point in the plane with the coordinates x_1, x_2. Let us now imagine that all points which have equal total costs are joined using one curve. We hence obtain a family of curves in our plane. Each curve is characterised by being the geometric location of all production levels with the same total costs. We can describe it as an indifference cost curve. We must now focus more closely on the shape of such a curve. Therefore, for reasons of simplification, we shall assume that total costs are monotonical in the narrowest sense.

Firstly, we must make an important observation. No indifference cost curve can be intersected by another indifference cost curve because then a vector would be "stronger" than a greater price vector,[15] which is impossible because of monotonicity. Secondly, and in the narrowest sense because of monotonicity, each direction only has one point where total costs have a predetermined level. The indifference cost curves hence define their radius vector as a definite function of the direction, which can also be continuously differentiated due to the regular nature of the total cost function. Thirdly, among all vectors that are defined by such a curve as product vectors with the same total costs, two vectors cannot occur such that one of them is completely contained in the other, unless they are consistent with each other. All indifference cost vectors are equally "strong".[16] That is, two arbitrary radius vectors of an indifference cost curve are obtained so that if one has a larger first component

[14] Each of these families of curves must fulfil the requirement that its curves do not intersect each other. Otherwise the requirement for an invertible definite covariance matrix of points and their coordinates would not be fulfilled.

[15] See Chapter 1, footnote 22.

[16] x_1 is defined by the curve to be a function of x_2 and vice versa, and in point of fact, the following always applies:

$$\frac{dx_1}{dx_2} < 0; \ \frac{dx_2}{dx_1} < 0.$$

For example, where the indifference cost curve has the point k_1 on axis 1 and point k_2 on axis 2, it transforms entirely into a rectangle which has these corner points:

$$(0, 0), (k_1, 0), (k_1, k_2), (0, k_2).$$

Within this rectangle they must monotonically decrease, but they can follow an arbitrary path.

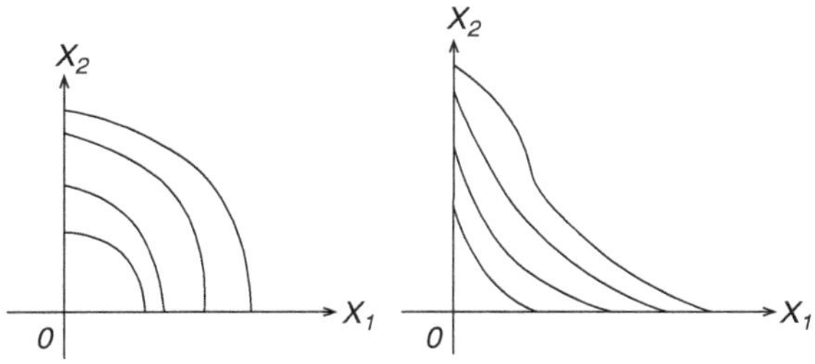

Fig. 3.4 Monotonically decreasing families of indifference cost curves

x_1 than the other, its second component x_2 must be smaller than the other's (see footnote 15). The curve therefore defines x_1 as a monotonically decreasing function of x_2 and vice versa. A family of curves such as this would, for instance, have the shape shown in the two graphs in Fig. 3.4.

The indifference cost curves can hence be either concave downwards or concave upwards.

These curves fulfil the requirement that radius vectors be defined as definite functions of direction. Similar laws to those that apply along concentric circles then apply along these curves. In combination with the family of radii through the origin, they result in a curvilinear system of coordinates.

We now look at the revenue function. We can construct "indifference revenue" curves for the revenue function combining all production levels that provide the same returns. These curves can have a very different shape.[17]

If we lay the family of indifference cost curves and the family of indifference revenue curves on top of each other, each indifference cost curve is therefore usually intersected by an infinite number of indifference revenue curves and vice versa. An innermost indifference cost curve which has a common point with the indifference revenue curve now corresponds to each indifference revenue curve. This point is then the production level which a particular revenue achieves with the lowest possible total costs. The geometric location of all of these points forms a curve. It is identical to the curve which occurs if we combine the points on the indifference cost curve which correlate respectively to the highest possible revenues. We refer to this curve as the curve of the most favourable directions and we can see that we do not allow the direction to vary along a circle but instead along the indifference cost curve. The most favourable production level point must also lie on this curve. Moreover, since the most favourable production level point must lie on the curve for the most favourable velocities of production, we hence obtain it as the intersection point of the curves of the most favourable directions and the most favourable lengths of production s [φ].

[17] For example, they can be closed curves which are located around a point, the maximum point (Translator's note: see Cassel, G. (1923): Book 2, Chapter VII, § 29: 267) of the revenue function.

In particular, where free competition is given, the indifference revenue curves appear to be parallel lines, cutting the vertical and horizontal axes into parts, and conversely behaving like the corresponding prices.

The points on the most favourable directions curve emerge as the tangential points of these lines and the indifference cost curves, and where incidentally the indifference cost curve must pass between the lines and the origin. Where a line has no such tangential point with a curve, the corresponding most favourable directional point lies on one of the two axes. If the indifference cost curves are now concave upwards, no tangential point of the kind referred to above can occur for any price combination.[18] The following proposition results:

(XXXII). Where the cost indifference curves are concave upwards, joint production can hence never occur under free competition. Instead, only good 1 or good 2 is produced (Fig. 3.5).

The accompanying diagram shows the construction[19] of the most favourable directions curve in the competitive economy.

This construction makes it possible to simplify a definition of the most favourable product vector as follows. By overlaying the two families of indifference curves, we obtain the connecting line for the tangential points as a geometric location for the most favourable production level required. This geometric location is also one-dimensional for more than two goods and is hence a spatial curve. The construction just described eliminates the unknown direction of the most favourable product vector from the problem. The most favourable direction curve defines direction as being a definite function of length. If we observe the supplementary proposition[20] for proposition XXVIII, we can then see from this construction that the general case of joint production is reduced to a special kind of tied proportion and through this, it is reduced to the case of the single production of a good on the basis of proposition XXVI. The most favourable length along the most favourable directions curve is determined according to the theory of single production. This can be achieved by examining the total cost function and the revenue function along that curve. However, the following simplification can also be made if we select one good, for example, good 1. The velocity of production for good 2 is defined as a function of the velocity of production for good 1 via the most favourable direction of the production curve. We therefore regard total costs and the returns for both goods simply as functions of the velocity of production for good 1. As a result, we have the general case for "single" supply and act according to the established rules.

[18] Here the tangential indifference revenue line would lie between the indifference cost curve and the origin, that is, in exact opposition to the claim put forward.

[19] The method for indifference curves was first used in theoretical economics by Francis Ysidro Edgeworth and, with groundbreaking success, by Vilfredo Pareto (1927): 540, footnote 1).

[20] See German ed. pp. 63.

Fig. 3.5 Upwardly concave indifference cost curves

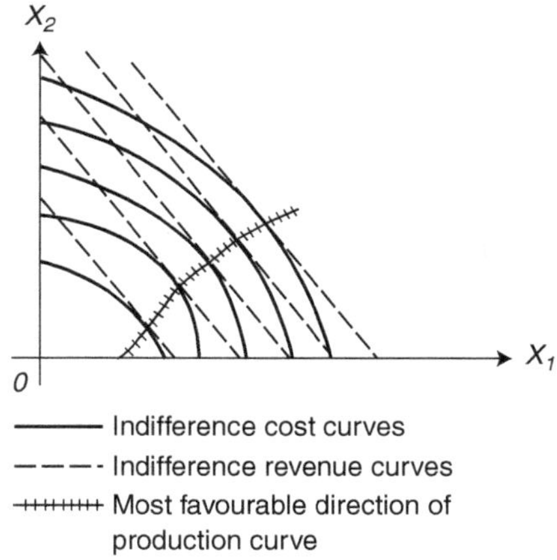

§ 3. Costs as an Untransformed Function of Two Velocities of Production

I.

In the first two sections we chose a detour via polar coordinates to present the theory of joint production because this was the only way we could prove that sophisticated notions about single production are also valid for joint production. This detour also allowed us to describe the total cost function more easily and thoroughly in the case of joint production than would otherwise have been possible. However, it is occasionally more beneficial to study joint production and its dependency on the velocities of production for individual goods based on a right-angled system of coordinates. This will particularly be the case in the next section where we will look at the theory of the inter-firm transfer price. In this section we shall investigate costs and returns as functions of the velocities of production for good 1 and good 2.

II.

1. Here, the total cost function is $K = K\ (x_1, x_2)$. We do not need to provide a description of its features because this has already been discussed in the previous two sections.

Only one question now remains and that is whether the total costs of joint production can normally be apportioned between the individual velocities of production in a meaningful way. For a case where goods are produced in a constant ratio, we have answered this question above in the negative (see Chapter 3, footnote 8). The same applies in principle to the general case. Firstly, it is always impossible to apportion constant costs in a logical way. The following applies to variable costs:

The variable costs for a particular velocity of production must fulfil two requirements. They must show how much is saved if this velocity of production is not obtained and they must state how much must be sacrificed to obtain this velocity of production. It is only if variable costs can be allocated to the two goods that these two requirements are clearly fulfilled and the two parts can be regarded as being the variable costs associated with each velocity of production. It is possible to show that such a definite classification of variable costs is then only possible if the variable cost function appears as the sum of two functions, only one of which depends on x_1 while the other only depends on x_2.[21]

2. We define the partial derivatives of the total cost function, that is, the marginal cost functions of good 1 and good 2, as x_1 and x_2 respectively. We denote them using K'_1 and K'_2. Both marginal cost functions are usually dependent on x_1 as well as x_2. In accordance with both variables we denote the derivatives of these functions as being K''_{11}, K''_{12}, K''_{21} and K''_{22}, where the equation $K''_{12} = K''_{21}$ exists because of the regularity of the total cost function. We refer to K''_{11} as the marginal cost gradient of good 1 and K''_{22} as the marginal cost gradient of good 2.

Average functions could be defined in a similar way but are meaningless here.

III.

In the case of free competition, the revenue function appears in the following structure:

---

[21] Variable costs for x_1 and x_2 are: $K_{II}(x_1, x_2)$.

Where x_1 is not obtained, the following variable costs occur: $K_{II}(0, x_2)$. Hence the following saving is made:

$$K_{II}(x_1, x_2) - K_{II}(0, x_2).$$

Where x_2 is not obtained, a saving of

$$K_{II}(x_1, x_2) - K_{II}(x_1, 0)$$

is hence made. If neither x_1 nor x_2 is obtained, then $K_{II}(x_1, x_2)$ is saved. However, it is now usually

$$K_{II}(x_1, x_2) \neq K_{II}(x_1, x_2) - K_{II}(0, x_2) + K_{II}(x_1, x_2) - K_{II}(x_1, 0),$$

that is, it is usually:

$$K_{II}(x_1, x_2) \neq K_{II}(x_1, 0) + K_{II}(0, x_2).$$

The equation exists only if $K_{II}(x_1, x_2)$ can be shown to be the sum of the functions of only one of the variables, when we can then write:

$$K_{II}(x_1, x_2) = K_{II,1}(x_1) + K_{II,2}(x_2).$$

Therefore the allocation of variable costs to individual goods is only possible in this case.

$$E(x_1, x_2) = x_1 . P_1 + x_2 . P_2;$$

In the case of a monopoly, we can generally distinguish where the price of a good depends on the supply of both goods, and in a special case, where the price of a good only depends on its supply.

We define the terms "marginal return" and "marginal return gradient" in the same way as the corresponding terms for costs. In the case of free competition, the marginal return of each good is its price and the marginal return gradient is equal to zero.

The description of the profit function similarly follows.

IV.

We look to determine the most favourable production vector in the case of rivalry, where we denote the components of this vector using s_1 or s_2. Profit now becomes maximised if these two equations $P_1 = K'_1(s_1, s_2)$; $P_2 = K'_2(s_1, s_2)$ are fulfilled and this results in a simple differentiation[22] of the profit function. We hence arrive at two equations, from which we can calculate the two unknowns, s_1 and s_2. These equations contain the following proposition which is copied from the corresponding proposition for single production:

(XXXIII). Where joint production and free competition exist through the application of the profit-making principle, each good then has marginal costs at the most favourable production level and these are equal to the corresponding prices.

A further condition for maximising profit is that marginal profit gradients are negative.[23] This results in a similar proposition to the corresponding proposition for "single" supply:

---

[22] We differentiate:

$$G(x_1, x_2) = x_1 . P_1 + x_2 . P_2 - K(x_1, x_2)$$

using x_1 and x_2 and obtain

$$\frac{\partial G}{\partial x_1} = P_1 - K'_1; \quad \frac{\partial G}{\partial x_2} = P_2 - K'_2.$$

Both derivatives of G must vanish for $x_1 = s_1$ and $x_2 = s_2$.

[23] It must follow that:

$$\frac{\partial^2 G}{\partial x^2_1} = -K''_{11} < 0$$

and

$$\frac{\partial^2 G}{\partial x^2_2} = -K''_{22} < 0$$

$$K''_{11} > 0 \text{ and } K''_{22} < 0.$$

hence:

(XXXIV). Under the conditions of proposition XXXIII, marginal cost gradients are positive at the most favourable production level.

The consequences arising from this do not need to be especially formulated because they already usually arise from an examination of the polar coordinates.

The third condition for maximising profit is that the Hessian matrix for the profit function is a definite and positive matrix. However, this condition cannot continue to be used for economic theory.

V.

We now look to determine the most favourable production level in the case of a monopoly, that is, in paragraph (a) for the general case and paragraph (b) for the special case.

(a) $$G(x_1, x_2) = x_1 . P_1(x_1, x_2) + x_2 . P_2(x_1, x_2) - K(x_1, x_2)$$
$$G'_1 = x_1 . P'_{1,1} + P_1 + x_2 . P'_{2,1} - K'_1$$
$$G'_2 = x_1 . P'_{1,2} + P_2 + x_2 . P'_{2,2} - K'_2.$$

For the most favourable production level, the following applies:

$$K'_1 = P_1 + s_1 . P'_{1,1} + s_2 . P'_{2,1}$$
$$K'_2 = P_2 + s_1 . P'_{1,2} + s_2 . P'_{2,2}.$$

$P'_{1,1}$ and $P'_{2,2}$ are certainly negative. What emerged from our explanations in Chap. 1, 2, 4. was whether $P'_{1,2}$ and $P'_{2,1}$ were negative or positive. Where two goods 1 and 2 are competing, both partial derivatives are hence negative and the marginal costs of the most favourable velocity of production in this most general example of a monopoly are certainly lower than the price. Where the two goods complement each other, both derivatives are positive and we are unable to decide whether marginal costs are higher or lower than the price. If we look at the various possibilities for the elasticity of demand we arrive at similar conclusions for the case where the two goods are neither competing nor complementary. In general terms, where a partial derivative $P'_{1,2}$ or $P'_{2,1}$ is negative, the monopoly price P_1 or P_2 is higher than the marginal costs for the most favourable velocity of production. Where a partial derivative $P'_{1,2}$ or $P'_{2,1}$ is positive, we cannot assert anything about the difference between price and marginal costs.

(b) In the special case of a monopoly, the following is true:

$$G(x_1, x_2) = x_1 . P_1(x_1) + x_2 . P_2(x_2) - K(x_1, x_2)$$
$$G'_1 = P_1(x_1) + x_1 . P'_1(x_1) - K'_1$$
$$G'_2 = P_2(x_2) + x_2 . P_2(x_2) - K'_2.$$

Here we immediately see that the marginal costs of the most favourable production level are lower than the price.

(The theory presented in the first two sections is in no way modified by these conclusions, but merely supplemented by them instead.)

VI.

In conclusion, we will say a few words about the principle of the satisfaction of needs and wants. We have previously seen[24] that the case of joint production must always contain a subsidiary principle because it only contains an assertion about the length of production, rather than about the direction of production. This proposition does not of course apply in cases where quantities are already determined beforehand. Here, the principle of the satisfaction of needs and wants means that the price of the vector unit is equal to average costs[25] for the production level in question. Determining the price of individual goods 1 and 2 is not possible from the principle of the satisfaction of needs and wants.

§ 4. Theory of the Inter-Firm Transfer Price

I.

We now wish to consider a problem that is important both economically and commercially. We shall study the problem of the inter-firm transfer price but only in general terms and in principle.

Let us imagine a profit-making enterprise that has the following structure: two firms consisting of firm 1 and firm 2. Firm 1 produces good 1 and this is used by firm 2 as a factor of production, for example, good 1 is fuel. Firm 2 produces good 2 which it offers in the market place. Generally speaking, this is now a question of the production of two goods by one enterprise, hence a special kind of joint production. The two firms however appear to be completely independent. They could also be two independent enterprises that are separated by the market, where firm 1 would then perhaps be firm 2's supplier.

We shall furthermore assume that firm 2 pays a price to firm 1 for the product 1 it supplies. We will describe this price as the transfer price. In this way, firm 2 should

---

[24] See Chapter 3, footnote 12.

[25] Hence:

$$\frac{K(x_1, x_2)}{\sqrt{x^2{}_1 + x^2{}_2}}$$

achieve an independent cost calculation and it subsequently regulates its velocity of production on the basis of general laws governing costs. We can now ask ourselves how high this transfer price must logically be and the answer is the real purpose of this section. Consequently, there is an additional question linked to this. How is the velocity of production of firm 1 regulated? We shall obtain the answer to both of these questions step by step, beginning with some simplified assumptions.

II.

We firstly assume that product 1 is not marketable and that it can be neither bought nor sold. Firm 1 produces the exact amount of its product 1 that firm 2 orders from it. Firm 2 produces good 2 and brings it to the market place. As a result, the entire enterprise therefore appears to be a producer of good 2 alone. To determine how much of this good the entire enterprise must produce in order to obtain the greatest possible profit and therefore, to determine the most favourable velocity of production for the entire enterprise, we do not even need to consider dividing this enterprise into two firms. We can establish this enterprise's total costs as the producer of good 2 and calculate the most favourable velocity of production using the established method based on the fundamental proposition of the profit-making principle. This is the starting point. There is now a requirement to obtain the transfer price between firms 1 and 2 so that firm 2 exactly achieves velocity of production s and no other velocity, thus complying with the laws governing costs. This is because it would then achieve a different velocity of production and hence the targeted total profit (Translator's note: "total profit" see Peacock: 81) could not be maximised. A transfer price which influences firm 2's cost structure must therefore be described as incorrect if a different velocity of production is achieved compared to velocity of production s, the most favourable velocity for the entire enterprise. The entire enterprise's product is identical to firm 2's product. The entire enterprise is represented in the market via its relationship with firm 2. Since firm 2 should orientate itself according to the general laws governing costs, the fundamental proposition of the profit-making principle is hence revealed to be a determinant of the production level for this firm. If firm 2 achieves the most favourable velocity of production s, firm 2's marginal costs must be equal to the marginal return. However, as the marginal return for velocity of production s is equal to the marginal costs of the entire enterprise and, as furthermore, this requirement applies to any arbitrary revenue function which the enterprise could find itself facing, we hence arrive at the important conclusion that the marginal costs of the entire enterprise and the marginal costs of firm 2 must always equal each other. Therefore, the marginal cost functions of the entire enterprise and firm 2 are also identical. This means however that the total cost functions of the entire enterprise and firm 2 must be equivalent to each other up to an arbitrary constant (because the marginal cost functions are derivatives of the total cost functions).

Let us look at these two total cost functions more closely. They consist of exactly the same items, except for the costs of product 1 which firm 1 produces. Here, the total cost function of the entire enterprise contains the total cost function of firm 1, whereas the total cost function of firm 2 is obtained by multiplying the quantity produced by firm 1 by the transfer price at the same point. Our earlier conclusion is therefore reduced to the following proposition: the total costs of firm 1 and the product of firm 1's production of velocity multiplied by the transfer price must be equal to each other for all of firm 1's velocities of production, except for one arbitrary constant. We could set this constant to equal zero unless it changes the enterprise's total profit. Then the total costs of firm 1 are obtained by multiplying firm 1's velocity of production by the transfer price. We hence arrive at this important proposition:

(XXXV). In a closed enterprise, the transfer price that firm 1 takes into account for firm 2 is equal to the average costs of firm 1, provided that firm 1's product is not marketable. Expressed another way, firm 1 supplies firm 2 according to the principle of the satisfaction of needs and wants.

Supplementary proposition: the profit-making principle remains applicable in terms of the entire enterprise if we add the quotient of an arbitrary positive or negative constant to the transfer price and firm 1's velocity of production.

The velocity of production achieved by firm 1 is also determined simultaneously. It is identical to the quantity of product 1 ordered by firm 2 in the unit time.

III.

We now change one assumption. We shall assume that product 1 is also marketable and that this product can also be bought and sold. The situation becomes more complex using this assumption. To simplify things, let us assume that free competition prevails.

First of all we shall examine the conclusion of the previous section more closely. There, we saw that the entire profit, except for a constant, must be assigned to firm 2 and that this firm therefore achieves the most favourable velocity of production for the entire enterprise. This principle is generally applicable. In reality, the total profit of an enterprise must always appear as a unit.

The following proposition applies to any profit-making production:

(XXXVI). The most favourable velocities of production for two enterprises are then equal to each other if their profit functions only differ by an independent variable for the velocities of production, that is, they differ by a constant.

This means, as it applies to the problem dealt with here, that the profit functions of firm 2 and the entire enterprise may only vary by a constant and hence the greatest profit that firm 1 may obtain is independent of production 2. In any case, the economic principle is preserved if the transfer price is set so that firm 1 obtains hardly any profit. This transfer price, which would represent a special solution for our problem, may be modified so that firm 1 obtains a constant profit. In the following pages, we shall firstly investigate the special solution we mentioned. The general solution is then revealed by simply broadening our study.

We therefore wish to observe how things could turn out if both goods were marketable. In this case, the quantities of product 2 produced by firm 1 in the unit time and the quantities of product 2 ordered by firm 2 in the same unit time do not need to be equal. Therefore we will denote the quantity firm 1 produces in the unit time using x_1, the quantity firm 2 orders using y and the quantity firm 2 produces using x_2.

The first fundamental observation is that firm 1 in our example (that is, with free competition and the marketability of good 1) is completely independent of the velocity of production it achieves and is therefore also completely independent of x_1 for firm 2 and the velocity of production y required by firm 2. Firm 1 must always achieve the velocity of production whose marginal costs are equal to price P_1 of its good. We denote this velocity of production using s_1.

Where y is actually less than s_1, firm 1 then obtains a profit by the additional production of $s_1 - y$, unless firm 2's profit changes. Where y is greater than s_1, firm 1 then has lower costs if it purchases $y - s_1$ at s_1 than if it produces the difference itself. From this it follows that the total costs of firm 1, which we denote using $K_1\ (x_1)$, always amount to $K_1\ (s_1)$, irrespective of y. From here, revenue flows out from the sale of $s_1 - y$ or the purchase amount flows in for the purchase of $y - s_1$, $(y - s_1)$ in both cases. P_1 is then added. The balance of firm 1 before paying y therefore amounts to:

$$K_1\ (s_1) + (y - s_1)\ .\ P_1.$$

This balance can be either positive or negative. In the first case, it implies costs and in the second, profits. For enterprise 1 to remain without profit or loss, the amount which firm 1 has to take into account for firm 2 and which arises as a product of transfer price V and the quantity ordered y, must be equal to the balance. The transfer price can therefore either be positive, that is, it must be paid by firm 2 to firm 1, or negative, that is, it must be paid by firm 1 to firm 2.

As a result, we derive:

$$y\ .\ V = K_1\ (s_1) + (y - s_1)\ .\ P_1.$$

Now this is not the only possible solution for determining transfer price V. We know that firm 2's profit can be distinguished from the profit of the entire enterprise by a constant. That is, a constant profit may appear in firm 1's calculation. Such a constant profit, one that is independent of y and x_2, is the amount firm 1 would gain (or lose) if it sold the whole quantity s_1 in the market place, that is, $s_1 .\ P_1 - K_1\ (s_1)$. We add this profit to the amount that firm 1 takes into account for firm 2. We then derive:

$$y\ .\ V = K_1\ (s_1) + y\ .\ P_1 - s_1\ P_1 + s_1\ .\ P_1 - K_1\ (s_1) = y\ .\ P_1.$$
$$V = P_1.$$

From this section, we arrive at the following proposition:

(XXXVII). Where firm 1's good is marketable and free competition prevails in its market, the transfer price is then equal to the market price. This is the special solution to the problem; the general solution emerges from the following supplementary proposition. The transfer price can be distinguished from the market price by the quotient of an arbitrary constant and the quantity ordered in the unit time by firm 2.

We therefore see that, under the assumptions made, it is immaterial whether the two firms are either tied to each other or separated by the market.

IV.

Where a monopoly prevails in the market for product 1 (we will consider the special case where price depends only on the quantity of its good), firm 1's market profit is hence:

$$(s_1 - y) \,.\, P_1\,(s_1 - y) - K_1\,(s_1)$$

whereby s_1 results as a function of y from the requirement that this profit is maximised, that is, from this equation:

$$(s_1 - y).\,P'_1(s_1 - y) + P_1\,(s_1 - y) - K'_1(s_1) = 0.$$

Now s_1 depends on y, that is, it is usually different from the velocity of production which would be achieved if firm 1 were acting independently. Since market profit also depends on y, the transfer price also depends on y.

The following applies:

$$\begin{aligned} y\,.\,V &= K_1(s_1) - (s_1 - y).\,P_1(s_1 - y) \\ &= K_1(s_1) - s_1.\,P_1(s_1 - y) + y\,.\,P_1(s_1 - y) \\ V &= P_1(s_1 - y) - \frac{s_1.\,P_1(s_1 - y) - K_1(s_1)}{y}. \end{aligned}$$

$P_1\,(s_1{-}y)$ is the market price of good 1. It is greater than transfer price V by the amount of

$$\frac{s_1\,.\,P_1\,(s_1 - y) - K_1\,(s_1)}{y}.$$

The numerator of this expression depends on y, and apart from in the case of free competition, it is not possible to bring the transfer price into line with the market price set as a constant profit for firm 1. As a result, we arrive at a fundamentally different outcome in the market for good 1 under monopoly in comparison to the outcome under competition. For a monopoly, the transfer price is fundamentally different to the market price.

This conclusion leads to an important insight. If the transfer price was also essentially equal to the market price for the monopoly, firm 1 could produce in conjunction with firm 2 as if it were not one enterprise, but instead had an independent economic interest in firm 2. Since this is not the case, it implies that the profit of the entire enterprise and hence the profit that is in the interest of the two joint firms is greater than the sum of their profits if they were independent of each other. We can formulate this proposition:

(XXXVIII). Where good 1 is the factor of production of good 2 and where the market for good 1 is monopolised, the total profit of an enterprise that produces good 1 and an enterprise that produces good 2 is greater if both enterprises form an entire enterprise rather than if they are independent of each other.

V.

For the case where firm 1 produces multiple goods, the similarity to the case for "single" supply can be established immediately on the basis of the theory of joint supply. So long as these goods are not marketable, a transfer price can only be constructed in the same way that it was constructed for average costs. Where goods are marketable, their transfer prices are hence equal to the corresponding market price in the case of free competition.[26]

References

Fanno, M. (1914). "Contributo alla teoria dell'offerta a costi congiunti" (Published in English as: A contribution to the theory of supply at joint cost, Forewords by Punzo, L. and Morishima, M., London, Macmillan, 1999) in Supplemento al Giornale degli Economisti e Rivista di Statistica, October, Athenaeum, Rome, via Calamatta, 16, p. 142.

Pareto, V. (1927). *Manual of political economy*, Paris (2nd ed.). (p. 540).

[26] The rather complex issues in this section are necessary to provide an accurate description of these mathematical notions and their origins.

Chapter 4
The Movement of Costs and the Structure of the Market Economy

The shape of the cost curve is a major regulatory factor for the production of an enterprise and because the sum of all of the enterprises represents the production of the market economy, the shape of the cost curve is hence an important construction component of the socio-economic system, together with the resulting laws. The purpose of this concluding chapter is to demonstrate its importance.

The theory of joint production, especially proposition XXX, enables us to use the notion of single production as a basis for all branches of production in a market economy. The fact that joint production exists does not change any statements about the market economy which are based on the assumption of single production. These statements only need to be discussed in greater detail on the basis of the above proposition. This conclusion enables us to greatly simplify our current examination. We only need to explicitly explore this problem of joint production where the direction of production immediately becomes the subject of the theory.

The complexity of the subject discussed makes it necessary for us to use isolation methodology. We therefore firstly describe the situation in the "static economy" by defining the term "static" very narrowly as in the sense used by Marshall.[1] The three factors of production: the technological level (Translator's note: for "technology" level rather than "technical" see Baumgärtner: 513), the state of the population and the level of needs are assumed to be constant here. At the start of this examination, we also assume the number and size of enterprises to be a given. We then investigate the transformation which this economy undergoes through the forces that are inherently generated, especially changes in the size and number of enterprises which lead to the "stable static". The treatment of the inherent tendency to concentration (Translator's note: "tendency to concentration" see Cassel: Book 1, Chapter III, § 14) forms the basis for the conclusion of the first section. In the second section, an approximate picture of the dynamic economy is derived from the static data by changing the data originally assumed to be constant, where the main

[1] Marshall, A. ibid.: 369–371: "stationärer Staat" (Translator's note: "stationary state" see Marshall, A.: 223). See also Clark, J.B. (1922): 132: "Imaginary Static Society".

H. von Stackelberg, *Foundations of a Pure Cost Theory*,

DOI 10.1007/978-3-642-34537-1_4,

emphasis is on the change in technology as the size of the firm increases. The third section is devoted to the influence of technological progress with the increased size of the firm (Translator's note: "size of the firm" see Peacock: 56, although von Stackelberg seems to apply this term to firms as well as enterprises).

§ 1. Regulation of the Static Economy

I.

In the second chapter we saw how a fixed general price level determines an enterprise's reactions. Firstly, the total cost function and the revenue function of the enterprise are fixed.[2] The fundamental proposition of profit-making production – as it applies to the competitive economy, that is, proposition XVI – determines the most favourable velocity of production and, consequently, the enterprise's supply,[3] which is achieved using a well-defined input vector.[4] This is merely demand generated by the enterprise's notional price level. As a result, the price level determines the supply and demand of each enterprise. We use the term "price" very broadly – in line with Cassel – adding wages, interest rates and pensions. The price level also determines the "individual equilibrium" for each consumer and therefore their supply and demand. If we total up the quantities demanded of any good across the number of demanding enterprises and individuals, we arrive at the aggregate demand for this good in the social economy we are considering. We can combine the total quantity of goods demanded to form the "demand vector".[5] In a very similar way, we arrive at the "supply vector". Both either depend on the current notional price level or, using mathematical terminology, on the current notional price vector.[6]

For an arbitrarily notional price vector, the corresponding supply and demand vectors would usually be different. However, competition law brings about the occurrence of a price level at which supply and demand are equal to each other – we can say this based on a reliable knowledge of all modern economic theory without giving detailed reasons. By equating the supply and demand vectors, we therefore arrive at a mathematical definition of the equilibrium price vector. However, this definition is clear only up to a proportionality factor. All equal price vectors up to a proportionality factor have the same supply vector and the same demand vector. This proportionality factor determines the purchasing power of the money unit. It is governed by the creation of money.

---

[2] See Chap. 1, § 1, III.

[3] See Chap. 2, § 4, at the end of II.

[4] Instead of all these, see: Pareto, V. *Manual of Political Economy*, Chap. III, especially No. 106–133.

[5] See Chap. 1, § 1, II.

[6] See Chap. 1, § 2, IV.

Equating the two vectors mentioned above is the equilibrium condition for our socio-economic structure. In addition to price relationships in a narrower sense, all income, all individually demanded and supplied quantities, all profits and the entire economic situation for every enterprise and every person in general are determined by the equilibrium condition. This situation for the individual enterprise is very different under our assumption of a coincidental size and number. One part of the individual enterprise does not produce anything at all, but instead lies idle because the enterprise's "minimum prices" lie above the market price for its products. We can rule this type of enterprise out of our investigation. This type of enterprise is only important to the total economy (Translator's note: "total economy" see Cassel: 18) for their operating equipment and can therefore be designated as "stocks of consumable goods" (Translator's note: "consumable goods" see Cassel: 56). Another part of this enterprise does produce, but finds itself in cost degression and incurs losses. The "quasi-rent" (in the Marshall sense (Translator's note: see Marshall: 74)) which they generate is not sufficient to meet constant costs. A third part ultimately finds itself in cost progression and achieves above normal profits and due to particularly advanced technological equipment, a particularly favourable location or particularly efficient management, etc., these enterprises' optimal price lies below the market price, that is, their "quasi-rent" exceeds constant costs. Within the last two groups outlined above, there can also be enterprises whose most favourable velocity of production accidentally coincides with their optimum.

II.

We now abandon the assumption that the number and size of enterprises is determined by outside forces, that is, as a starting point we can probably assume a certain distribution of enterprises which is influenced by economic forces and resembles our socio-economic structure. In this case, we can state that the equilibrium described a moment ago is only an equilibrium "in the short run" (Translator's note: von Stackelberg's original text has this phrase in both English and German. We have omitted the repetition). At the very moment we have freed our system from the "shackles" of a constant distribution of the production setup, we introduce a redistribution process, firstly within each branch of production and secondly between the branches of production. Economic entities are motivated by the pursuit of profit.

Within the individual branches of production, each enterprise begins to strive towards improving its production setup thereby increasing its "internal and external economies" (Marshall (Translator's note: see Appendix H, footnote 82)) in order to increase its profit. Enterprises aim to achieve the best possible production setup under the given technological and social conditions. Old machines are replaced by new ones, new production processes are introduced, and capital is withdrawn from poorly located enterprises and reinvested in those which are better located. Branches of production with poor profit maximisation are abandoned and enterprises try to identify those where the outlook is better. Enterprises fail if they cannot keep pace with this conversion process. Population migration is accompanied by capital reallocation and so a wide-ranging and complex conversion process occurs and this leads to the establishment of a lasting general equilibrium, a general equilibrium "in the long run" (Translator's note: von Stackelberg's original text has this phrase, "in the long run" in English).

To demonstrate the process's main features, we return to using isolation methodology. Firstly, we rule out a redistribution of land, work and capital between the branches of production and assume that the individual factors of production are homogenous. We regard entrepreneurial activity simply as a special form of work activity. In this case, the final result will be the development of a standardised proposition within each branch of production for pensions, interest and wages, including the entrepreneur's wages. There cannot be other types of income here. Each entrepreneur earns the same as each worker. All enterprises operate in exactly the same way as each other. They all have the highest possible technological level. Revenue can be completely divided into pensions, interest and wages. If, as we can clearly see is necessary, we regard these three types of income to be components of costs, it therefore follows that the total revenue for each enterprise is equal to its total costs for each branch of production. We can hence formulate this important proposition:

(XXXIX). In a static equilibrium, each enterprise finds itself in the optimum position.[7]

If we now abandon the isolation of the individual branch of production, then the factors of production can be relocated from one branch to another until the same proposition prevails everywhere for pensions, interest and wages as well as productively exploiting all the existing supply of the factors of production. What we said about the individual branch of production now applies uniformly to the whole of the social economy.

The general price level now either determines a firm's most favourable structure or, as we can term it rather imprecisely using everyday language, the size of the firm. On the one hand, we can describe this using the amount spent on land, capital and work or on the other hand, we can use the corresponding optimum velocity of production. The price level further determines the demand for the product concerned. The quantity demanded in each unit time divided by the optimum velocity of production gives the number of enterprises and these also therefore depend on the price level. Since the number is always a whole number, our statements only apply if the quantity demanded is an integer multiple of the optimum velocity of production. Let us assume this to be very small relatively speaking, which we also have to assume as a necessary condition of "free competition". We can then accept the imprecision of a generalisation and make our statements without reservation. The requirement for work, capital and land by any branch of production is determined by the number of enterprises, which we

[7] See Marshall, A. ibid., Book V, Chap. 11, § 6.

assume to be a single firm, and by the fact that everything is produced in the optimum position. This requirement must be met by the existing supply of factors of production. If it is not, prices of scarce factors of production hence increase while prices of abundantly available factors decrease until the equilibrium is reached.

Finally, let us also abandon the assumptions of the homogeneity of capital goods. Then work, capital and land appear to be groups of factors of production which are ranked in terms of quality. The overall picture of the economy becomes more complex. The firms of a branch of production are no longer the same as each other but differ in terms of location, management quality, etc. These differences consequently have the effect of making real profits exist in a diverse variety of formats and these are awarded to the preferred enterprises. They result from the price differences of the qualitatively different factors of production and must be regarded as costs within our system because each individual factor of production is the subject of demand.[8] Our formal proposition (XXXIX) therefore remains intact.

This proposition leads to an interesting consequence for allocation theory. Each enterprise uses so much of each factor of production that the marginal productivity of this factor is equal to its price. If we regard the enterprise's return to be a function of the input vector, we can also say that each enterprise achieves an input vector where the gradient of its revenue function is equal to the price vector of the input. Besides, since total costs are equal to total revenue in the equilibrium position, the social income is entirely allocated on the basis of the factors of production and in accordance with the principle of marginal productivity.[9]

III.

We have now painted a picture of a static competitive economy with a number and size of firms that is inherently determined. We saw how the profit-making principle produces this requirement. However, we argued that each firm should be an independent enterprise. We also abandon this condition and ask ourselves whether there are forces in the system of the static competitive economy which promote the principle of largest possible decentralisation and which lead to concentration.

Such forces really do exist. They primarily have an effect on horizontal firm concentration and secondly on vertical firm concentration. However, this force is not an obligation in the system of free competition but is instead just a trend, a "temptation". Firstly, it shows that the enterprises in a branch of production are

---

[8] Instead of all of these, see also: Walras, L. (1926) 175 et seq.: "Des capitaux et des revenues" (Capital and Income).

[9] See Appendix A, VII. Similarly: Moore, H.L. (1929): 145. By contrast, see Aftalion, A. (1911): 346–369, "Les trois notions de la productivité et les revenus" (Three Notions of Productivity and Revenues). Revue d'Économie Politique, vol. XXV. Mayer, H. (1925). Article "Zurechnung" (Allocation) in: Handwörterbuch der Staatswissenschaften, especially I, 2, c.

capable of achieving a greater profit if they combine to form a monopoly. We know from proposition (XXI) that the monopoly price is always greater than marginal costs, while the competitive price is equal to them. The monopolist could however very probably also achieve a production level which is competitively the most favourable. If he does not do this, it is only because he can obtain a greater profit with another production level. Consequently, a tendency towards monopolistic concentration within the branch of production results from the profit-making principle. This tendency is not a "systematic obligation" because the competitive economy is also functional. The tendency to form a monopoly is only a side effect of the profit-making principle. This concentration of power collides with internal opposing powers so that the stronger these powers are, the greater the number of enterprises there are in a branch of production. They do not need to become generally accepted at all and can consequently be ignored at first, especially as the largest concentrations of power only occur in a dynamic economy. Furthermore, a decentralised power results from the established fact that the danger from outsiders can be a threat to every monopoly. As soon as a branch of production is monopolised, its prices provide greater profit maximisation possibilities compared with other branches of production and attract entrepreneurs into the monopolised branch of production in larger numbers.[10]

At first we do not see a tendency towards vertical concentration resulting from the system. We have seen (in proposition XXXVII) that the transfer price of a marketable commodity is equal to the competitive market price and that no "systematic" profit therefore occurs from a vertical merger of two firms within the competitive economy. However, where a branch of production is monopolised, this results in a tendency towards the incorporation of processing firms on the basis of proposition (XXXVIII). The vertical combination results from a horizontal concentration as a secondary phenomenon. Consequently, the same obstacles stand in the way of this as they do with horizontal concentration. We can hence say that the organisational structure of the static competitive economy is relatively stable. It does not contain any power derived from the system which would necessitate a fundamental transformation.

§ 2. The General Effects of Dynamic Changes

What our static economy particularly demonstrates is the constancy of the population, the factors of production and the technological level. By "loosening" these three "shackles", we arrive at an approximation of the dynamic economy.[11]

[10] See Barone, E. ibid. § 158: "Potentielle Konkurrenz" (Potential Competition).

[11] Clark, J.B. (1922): 203–206, distinguishes five types of dynamic influences: (1) increases in the population, (2) increases in capital, (3) changes in the production method, (4) changes in the enterprise's organisation and (5) changes in tastes. We do not take (5) into account for our investigation. We can combine (3) and (4) using the term, the "change in the technological level". We separate (1) into "change in the population", which we firstly only consider to be a change in the demand side, and "change in available work".

We return to using isolation methodology. First we abolish the population constant, then the individual factors of production constant, and finally, the technology constant.

I.

A change in the population size can—ceteris paribus—quickly be dealt with within our set of tasks. It will only mean a shift in the composition of the goods produced. An increase in the population implies a relative shortage of supply. As a result, goods with inelastic[12] demand will be produced in greater quantities than before, whereas those with elastic demand will be produced in smaller quantities[13] than before. A decline in the population has exactly the opposite effect. It is obvious that the full effect of these two tendencies only really becomes clear "in the long run" because it takes time to dismantle existing production facilities, which will be necessary in both cases.

II.

Land is a factor of production that we can keep constant. First of all, it is enough to vary two factors of production in order to be able to identify all the important consequences. Conversely, we must essentially consider land to be an unchanging variable for a given market economy. We will therefore only look at the variations in capital and work.

Every change in the existing supply of capital or work shifts the proportion of factors of production, resulting in a structural change in the optimum firm. A redistribution process occurs in every enterprise which however, can only have an effect "in the long run". At the same time, the market economy's supply situation changes—assuming a constant population—which leads to a redistribution process between the branches of production, depending on the elasticity of demand for individual goods.

1. An increase in capital results in a better opportunity for each enterprise to provide itself with capital than before. In the short run, this does not firstly imply any change in the organisation of indirect factors of production or capital goods but simply implies an increase in short term capital investment. In time however, a general "rationalisation" is carried out in each enterprise with regard to capital intensive production methods. The ultimate factor of production combination

---

[12] Elasticity < 1.

[13] Elasticity < 1.

enables the market economy to be as abundantly supplied with goods as before, which conversely means an increase in production compared with the situation before the increase in the capital available.

The special characteristic of each enterprise, as we explained earlier in detail, is the fact that it achieves its optimum position in a long-run equilibrium. Here however, for each factor of production—we can easily show this[14]—the law of diminishing returns applies, that is, marginal productivities decrease with the increasing quantity of each enterprise's factors of production. This applies both "in the long run", and definitely "in the short run". On the other hand, we should assume that the marginal productivity of a factor usually increases with an increasing quantity of another factor of production,[15] and indeed in a similar way "in the short run" and "in the long run", although in this latter case, a greater increase occurs. For our case, it follows that to begin with, each enterprise progressively arrives at a greater under-use of capital. The marginal productivity of capital decreases and this increases the amount of work. Overall, there is a decrease in real interest (Translator's note: "real interest" see Cassel: 109) and an increase in real wages (Translator's note: "real wages" see Cassel: 321) so the enterprise makes exceptional profits. In the long run, each enterprise arrives at a new optimum position usually by intensifying capital with a simultaneous intensification of competition. The marginal productivity of capital (and, as a result, of real interest) increases once again but not to the original level. The marginal productivity of work, and as a result, of real wages, usually also increases. Exceptional profits are absorbed. The abundant supply possibility for the whole market economy requires a redistribution between the individual branches of production. Here, goods with elastic demand are produced in proportionately greater quantities than goods with inelastic demand. In absolute terms, all types of goods are produced in larger quantities than before. However, we cannot decide whether the number of enterprises in the individual branches of production has become either greater or smaller because we are unable to determine whether the new optimum quantities supplied are either greater or smaller than the old ones without a detailed knowledge of the individual production functions. The relative share of the social product (Translator's note: "social product" see Peacock: 275) allotted to work in the competitive allocation process, has grown. Assessing whether a capitalist's real income (Translator's note: "real income" see Peacock: 247) has either increased or decreased in absolute terms, depends on the long run elasticity of demand for capital in the market economy, which is re-determined by the technological level.

2. An increase in available labour shows very similar phenomena to the increase in capital, mutatis mutandis. We can immediately combine these results. There is a redistribution of the factors of production from the branches of production with an

[14] It results from the second minimum condition for average costs.

[15] The reason for this fact is that a complementarity relationship (Translator's note: see Peacock, A.: 65) between the factors of production is stronger than if it were a substitution relationship (Translator's note: ibid.). Investigating this more widely would be too much work here. In addition, see Chapter 1, footnote 25.

inelastic want towards those with an elastic demand. In the short run, individual enterprises leave their optimum position and arrive at cost progression. They make exceptional profits, particularly enterprises which produce for an elastic demand. As real interest increases, real wages decrease. In the long run, enterprises rationalise in the direction of more labour-intensive production methods by converting their operating equipment too. They therefore arrive at a new optimum position as a result of competitive pressure. Exceptional profits are absorbed. Real wages increase, but do not return to the initial level. Real interest probably undergoes a further increase. The market economy's supply has become more abundant overall.[16] We cannot determine the extent of the effect on the increase in the social product and work, or how the number of firms has changed unless we have more detailed information.

3. A decrease in existing capital operates in a similar way to an increase in the work available, but with the following modifications. Individual enterprises initially arrive at cost degression and incur losses. The market economy's total supply becomes scarcer. There is redistribution in the direction of a relatively abundant satisfaction of inelastic wants. Unquestionably, the market economy's supply has also become scarcer after carrying out a transformation in the long run. With respect to the increase in work, a significant difference is that interest firstly increases but is likely to decrease during long term redistribution, although not as far as the initial level.

4. We can easily estimate a decrease in the labour available based on the effects outlined above.[17]

---

[16] Exceptional profits (or exceptional losses) occur whenever the enterprise arrives at its optimum position in cost progression (or cost degression). Here, the entrepreneur's real income is greater (or smaller) than his "normal income in the long run" (Marshall). Exceptional profits are a residue. In this case, the principle of marginal productivity does not produce a full allocation. The absorption of exceptional profits does not for instance imply a decrease in the entrepreneur's income; it just brings it into line with the normal level in the long run (in certain circumstances due to rationalisation, this income actually increases by an amount that exceeds the amount of exceptional profits).

[17] We will now add two remarks on the issue of generalisations about the earlier results in this section:

(a) We have primarily carried out our critical analysis as if there were only two factors of production, that is, capital and work. The results can be generalised without difficulty if we include land and entrepreneurial services as factors of production in our investigation. The assumption that is also more complicated but closest to reality, that is, the inhomogenous composition of the individual factors of production, leads to observations that are fundamentally the same. However, we cannot trace this idea back. That would require a specific investigation.

(b) Mathematically speaking, all functions that occur in our theoretical concept are "location functions". Therefore, we can only use isolation methodology if it is a question of comparing a starting and finishing level and then studying the effects of individual changes in turn. We shall content ourselves with that here. An examination of the "methods" would last too long. We can arrive at the final outcome of a simultaneous increase in population and work, for example, by firstly studying a decrease in work and then in population size. The intermediary stages of our analysis would however deviate from reality. A closer examination must really be the subject of a special study.

III.

1. So far, we have concentrated on looking at the effects of quantitative changes from a mathematical point of view, that is, using the effect of changes in independent variables. The change in the technological situation appears by contrast to be a qualitative problem which is quantifiable for certain purposes. We no longer concentrate here on changes in variables but instead on changes in the functional relationship. Technological development adjusts the production function itself. We now have to concentrate on the effects that result from this.

Technological progress,[18] from an economic point of view, is an improvement in the market economy's supply possibilities. Compared to the starting position, it leads to the same amount of goods being obtained with a lower input of factors of production or a greater amount of goods being obtained with the same input of factors of production. It hence implies either a partial or general shift upwards of the production functions. Its general effect is to increase production and simultaneously cause a redistribution towards a relatively stronger supply of elastic need.

Technological progress can increase marginal productivity in the same proportion. This does not then change anything about the competitive allocation criteria of the social product. Usually however, progress only affects certain branches of production and changes the productivity of the factors of production to different extents. As a result, there are shifts which we shall now look at in detail.

If only one branch of production is affected by technological progress, it therefore implies that a greater production quantity can be produced with the previous input of work, capital and land. The reduction in the price of products is inversely proportional to the increase in the number of products. The ultimate impact differs depending on the elasticity of demand for the product concerned. If demand is inelastic, a share of the factors of production is transferred from the branch of production in question. The supply of other goods becomes more abundant. If the elasticity of demand is equal to 1, no shift in the factors of production then results at all. If demand is elastic, the rationalised branch of production attracts new factors of production to it and other branches of production lose their factors of production. The supply of other goods becomes scarcer.

The changes that cause technological progress in combination with the factors of production, do not produce[19] any change in the proportion in which work, capital and land are combined collectively within the entire social economy. This is because it is unthinkable that factors of production would only lie idle in a free

[18] The term "technological progress" should be understood very broadly in the following pages. By "technological progress" we mean the improvement in production methods. This technological progress then includes scientific-technological discoveries and inventions, as well as improvements in structure, location and supplementary apparatus, etc.

[19] If, for reasons of simplification and along with Cassel, G. we assume a supply of factors of production that is independent of price.

competitive economy in marginal cases. However, these changes lead to a change in marginal productivity which leads to a change in the socio-economic allocation criteria.

Each technological improvement can be interpreted as an increase in the factors of production. As a result, we can immediately evaluate these effects based on our earlier explanations. The invention of labour-saving machines means, for example, that the same product is produced with a smaller work input. As a result, work has become more abundant compared to capital and land. An increase in workers, which we discussed earlier, has the following effects: pressure on wages, redistributions, etc. The introduction of new land management methods, e.g. improved crop rotation, operates in the same way as an increase in land. The introduction of cheaper machinery with equal productivity, e.g. cheaper engines, is synonymous with an increase in capital. The effects are clear from the above explanations and do not need to be explained here again. We can see that the implications of technological progress explained a moment ago are not a new problem.

2. The question of the influence of technological progress on the size of the firm is more important to us. Due to the change in the production function, the geometric location of a full allocation point shifts, which, as we have shown, is a key feature of the optimum position. In consequence, the optimum for each direction of production shifts and so a fundamentally new situation then exists.

Technological progress can essentially have two effects on changes in the optimum firm size. It can either reduce or increase it, that is, the optimum quantity supplied can be either smaller or larger. If the optimum position becomes smaller, there is a tendency for new firms, which are particularly competitive compared with old enterprises, to be formed. As a result of this, there is therefore simultaneously an obligation for old firms to rationalise and to adjust to a smaller size. In the end, rationalisation will always be carried out everywhere. Since the optimum quantity supplied has decreased and total production (Translator's note: "total production" see Cassel (1967): 60) for the branch of production concerned has by contrast increased, the number of firms has increased overall. The tendency we mentioned earlier for new enterprises to become established therefore not only serves to intensify the obligation to rationalise, which is already fixed for them by the profit-making pursuit of profit, it also serves to correspondingly increase the number of existing firms. If we were to assume for instance that no new enterprises appeared at first, then even more new firms must indeed be added subsequently following the rationalisation of old firms. We can hence state that technological progress, which is linked to a reduction in the firm size, has an impact via a fairly smooth process within a given branch of production. In the first stage, enterprises partly have exceptional profits and partly have losses and these eventually balance out as rationalisation is completed. The elimination of firms as a result of technological progress usually only occurs when new firms appear in larger numbers than necessary. Overall, technological innovation produces a continuous upturn in this case.

Things are fundamentally different when technological progress is linked to an increase in the optimum position. Here too, technological innovation produces a tendency towards rationalising existing firms, which is strengthened by a tendency to set up modern firms. However, ultimately, if the increase in the optimum position is not insignificant, it results in a reduction in the number of firms in the branch of production concerned. Setting up and converting new firms initially occurs due to an upturn. A more rational production possibility leads to exceptional profits and results in enterprises racing each other. The decrease in prices caused by increased supply eventually absorbs profits. Even so, rationalisation is carried out right to the end because it at least leads to a minimum loss. Now however, there is an obligation to eliminate firms and this continues until the correct number is achieved. The price level decreases so that the weakest firms eventually have to lie idle and they must eliminate themselves from the production process. At first, this technological progress produces an upturn with an increase in the optimum position and this expands the firm, but then it leads to a downturn which restricts the number of firms.

In outline, this is the effect of technological progress on the number and size of firms. A more thorough investigation lies outside the scope of this book. However, there is one question which has emerged from the task dealt with just now which has certainly interested us and that is the effect of the growth in the optimum firm size on the socio-economic organisation of production insofar as this link exists based on formal laws governing costs. The concluding section is devoted to this question.

§ 3. The Influence of Technological Progress on the Economic Model

Whereas the decrease in the optimum firm size does not affect the competitive organisation of production, this is different with an increasing optimum position. Since the increase in the optimum position is a characteristic feature of our earlier economic growth, it is appropriate to discuss transforming the social economy's production setup which is associated with this in more detail. It goes without saying that we need a short theoretical preliminary discussion of the so-called "polypoly" to actually deal with this problem.

I.

Strictly speaking the following assumptions underpin our examination of the competitive economy:

1. The quantity supplied by the individual firm is infinitely small in practice compared with the total output by the branch of production concerned.

2. The number of rival enterprises competing with each other is correspondingly infinitely large in practice.
3. The price function is therefore virtually a constant for each firm.

Strictly speaking, these assumptions are incorrect in real life, but they represent real relationships more accurately when there is a large number of firms and hence enterprises. At the same time, they make it possible to greatly simplify our theoretical study. We can now ask the following question, "What profit-making regulation of production is there if the number of firms is fixed and is not particularly large?" The standard answer to this question can only be found using mathematical methods due to the complexity of the relationships and the links between firms. However we can initially be happy with the answer to the question in the case of only two rivals, that is, in the case of a so-called "duopoly" (Translator's note: in this work by v. Stackelberg (1932) "Duopol" is used in the German rather than "Dyopol" (see v. Stackelberg's Marktform und Gleichgewicht, 1934), therefore we have replaced "dyopoly" with "duopoly" for this translation). We can now refer to the theoretical work already completed and forgo our own deductions.

This is the "intermediary stage" as mentioned above in Chapter 2, footnote 42. Rather than citing everyone, we have chosen to only refer to Kurt Sting. He describes the problem simply and correctly in his essay, even if in our opinion he gives it undue weight. The basic idea is as follows. Let A and B be two producers of the same good. We firstly assume that each one treats the other's supply as a given variable which he must simply resign himself to. For each quantity supplied by B, A will then allocate a separate quantity supplied at which he makes the greatest possible profit and the same applies to B. For each, their supply is hence revealed as a function of the other's supply. If we regard the quantities supplied by A and B as the two unknowns, we hence obtain the two equations in the "reaction functions" and from these, we can clearly define the two unknowns. As a result, the economic equilibrium, described by Sting as "polypolitical pricing", seems to be fixed. However, this is not really the case. If we assume complete "market transparency", then there must exist, for example, a "hyperpolitical" pursuit (in Sting's words) for A. That is, A will not base his supply on his own function, but instead on B's reaction function because he then has greater profit maximisation options. Since he knows that B regards his (A's) supply to be a given variable, he will choose the combination that is the most favourable for him (A) from all the combinations which are possible for B on the basis of his (B's) reaction function. However, he will then usually obtain a different supply to the supply that matches his own reaction function. Now B tries to behave in the same way. This means that each tries to force his supply on the other as an inevitable fact. An equilibrium is impossible in such a situation. Each one tries to force the other back to his reaction function and so there is a fierce competitive battle, which must end either with the surrender or the destruction of the rival if an agreement is not reached. We therefore see that a duopoly does not lead to an equilibrium.[20]

[20] An interesting discussion about this problem has taken place in research articles. Cournot, A. was the first to tackle it and he made implicit assumptions that each of the two suppliers takes the respective supply of the other as a given variable, but he does not deal with the other possibility.

An equilibrium is possible, subject to the assumptions made (implicitly) by Cournot. Here, each enterprise consequently behaves as if it were unable to influence the supply of the other and each enterprise accepts the other's supply as a given. We shall describe this situation as the "Cournot duopoly". Furthermore, an equilibrium is possible if one of the two enterprises, for example A, takes the supply of the other (B) for granted as a given but enterprise B behaves "hyperpolitically", that is, the enterprise exploits A's compliance and achieves a supply that gives it the greatest profit in the given circumstances. An equilibrium is impossible however if both enterprises behave "hyperpolitically", a situation which we would like to describe as a "Pareto duopoly".

If the Pareto duopoly and the Cournot duopoly were the only two possibilities, we could then agree with Cournot's supporters that in practice, if participants assess the whole situation correctly, then only the Cournot duopoly is possible because the attempt to react "hyperpolitically" cannot lead to any reasonable result. Now in reality however, there is the possibility for each enterprise to act "hyperpolitically" to exploit their rivals. If A gives up any "hyperpolitical" attempt and makes do with a smaller profit, then from his point of view B does not have the slightest reason to exploit the "naivety" of his rival, to give up exploiting it or to increase his profit over and above the possibilities of the "Cournot duopoly". Theoretically speaking, there remains no other possibility than to deny the ability of the profit-making principle and produce a mechanical equilibrium in the case of the duopoly. The rivals must come to an agreement of some kind and they must supplement the inadequate economic mechanism in this situation using economic policy.

The principal difference between the Cournot and the Pareto duopoly continues to exist if we assume that there are not two, but instead multiple enterprises which compete with each other. We can then replace the word "duopoly" with the term "polypoly". We will leave to one side the interesting problem that occurs in this case. For us, the following statement is important: the greater the number of approximately equal enterprises, the smaller the difference between the Pareto polypoly and the Cournot polypoly, as well as between the Cournot polypoly and the situation under free competition. This becomes the additional profit which individual enterprises could achieve with a "hyperpolitical" reaction by comparison with the Cournot polypoly, and it becomes an increasingly smaller profit with an

---

The criticism by Bertrand, J., Edgeworth, F.Y., Marshall, A. and Pareto, V. is relevant here. This led to a fairly general rejection of Cournot's line of argument but without the question of the assumptions being checked in enough detail. Wicksell, K. and Schumpeter, J. are similarly not entirely correct about this matter of the Cournot solution. Similarly, Schneider, E. is not correct either. Amoroso, L. indeed sees the difference between the assumptions but ignores the potential and meaning of the competitive battle described above. Kurt Sting again deals with the entire problem very satisfactorily. He works clearly through the assumptions. However, he might be overestimating the practical importance of "polypolitical" pricing and the Cournot duopoly but underestimating the reality of "hyperpolitics". In the literature, see the article cited by Kurt Sting in Chapter 2, footnote 42, which provides a good overview of the literature. Again, see Schneider, E. (1932). (Beiträge zur ökonomischen Theorie (Contributions to Economic Theory): 4.)

increased number of rivals. The additional profit with the Cournot polypoly also decreases compared to free competition. Expressed another way, it eventually makes no difference to the enterprise whether it either attempts to influence the supply of its rivals and the price, or just the price, or neither of these. The difference in profit becomes so small that it cannot bring about any more changes to the system. At the same time, the difference between the quantities supplied becomes increasingly smaller in the Pareto polypoly,[21] the Cournot polypoly and in free competition. In this way the equilibrium is only achieved in practice in free competition if the number of enterprises is sufficiently large.

II.

1. The dynamically limited reduction in the number of enterprises of a branch of production must eventually lead to a point at which the problem of polypolies surfaces. Habit, confusion in the market, calculation difficulties, etc. may postpone this point at first. However, the problem surfaces at any time. It could well be that only the Cournot polypoly is possible at first because compared to free competition, its difference is perhaps greater than the difference under the Pareto polypoly in comparison with the Cournot polypoly. Then an equilibrium in fact exists, but the fundamental proposition of profit-making production is no longer fulfilled competitively and the price is now higher than the marginal costs for each enterprise. Conversely, since no branch of production in relation to any other, can make exceptional profits after restabilisation and since the price is hence equal to average costs, we therefore reach the interesting conclusion that a transformation of the enterprises from the optimum is linked with a transformation to a polypoly in the degression zone (Translator's note: "degression zone" see Peacock: 301). It is entirely possible that the individual demand curve, that is, the demand function which the individual enterprise is faced with in the polypoly, is tangential to the average cost curve at a point to the left of the optimum, and which incidentally follows a path below the average cost curve. Since the most favourable velocity of production for the Cournot polypoly is smaller than the optimum, the number of enterprises is determined in a different way here in comparison to the case of free competition.

Each enterprise obtains a profit that is greater than the profit they would have gained in the same branch of production if they had been faced with the same number of enterprises under free competition. By contrast, the polypoly profit is not greater than the profit the enterprises obtain in other, competitively organised branches of production. If that were the case, then the polypolistic branch of production would

[21] In the Pareto polypoly, each enterprise attempts to launch a particular quantity and to force it on the others as an unchangeable variable. The total of these quantities is the Pareto polypoly's aggregate supply which naturally involves heavy losses for all rivals and is therefore only able to be sustained for a short time.

attract entrepreneurs until profit maximisation options were the same everywhere. Conversely, some entrepreneurs would leave this branch of production if it was regulated by the competitive principle, rather than still regulated by the polypolistic principle. They would turn to other branches of production where the price of the entrepreneur's services, that is, the normal entrepreneur's income, would decrease in every respect. The transformation from absolute rivalry to polypoly then affects the whole social economy, such as in the case where there is a shortage in the entrepreneur's services. The more branches of production which become transformed into a polypoly, the greater the entrepreneur's share of the social product. This definitely applies to a monopoly. The profit-making principle has a less rational economic effect on a polypoly, and definitely on a monopoly, than in free competition, a conclusion that has already been the universal theory since Cournot.

2. A further reduction in the number of enterprises leads to a Pareto polypoly sooner or later. The profit-making mechanism now breaks down. There is an obligation to come to some sort of agreement, whether it is on an equal basis, which would initially equate to a cartel, or whether it is instead determined by the fundamental proposition of giving precedence and taking precedence, where one enterprise wins a victory over the other and "hyperpolitically" determines market conditions. A cartel is the most likely outcome if there is a large number of approximately equal enterprises. Such a cartel, however strict it is, eventually leads to the possibility of the monopolistic domination of the market and as a result, the competition mechanism ceases to regulate production. The market economy link between the firm size and the number of enterprises breaks down as the horizontal concentration becomes more tightly organised and in fact, is in a similar proportion. Ultimately, questions about the number and size of firms are completely dominated by opinions that exist within the organisation as a whole. These opinions can be influenced by external factors and be based on profitability calculations for the organisation as a whole. In this last case, enterprises will strive to achieve the most rational total production, but pursue monopolistically determined profits.

3. The dynamic process described a moment ago not only formally includes certain tendencies towards horizontal concentration but vertical concentration too. Under the Cournot polypoly and based on the profit-making principle, firms already strive for concentration, producing lower order goods because proposition (XXXVIII) shows that a vertical merger is advantageous in the case of a monopoly, that is, whenever the price level depends on the quantity supplied to individual enterprises. This vertical tendency to concentration becomes stronger the greater the progress that horizontal concentration makes. At first it is just a "temptation" resulting from the pursuit of profit rather than a systematic obligation. This kind of systematic obligation arises however if we assume that technological innovation leads to horizontal concentration in two branches of production which each follow the other. The resulting cartels cannot enter into a contract with each other on the basis of a mechanical general equilibrium. A power struggle would occur because each partner would try to force his price offer or conversely, his price demand, on the other. The result is the necessity of coming to an agreement and cartels are then systematically forced into vertical concentration.

III.

Another increase in the firm size now becomes a matter of internal horizontal concentration. The more widespread centralisation is, the more we can regard each branch of production as a single enormous enterprise, which in terms of its operational composition can be described as a "system of batteries connected in parallel". Here, each component firm is operating as an optimum firm. The crucial difference compared to the competitive economy is that the profit-making principle no longer guarantees economic productivity but opposes it instead.

The tendency towards stricter centralisation increases as the optimum firm size increases in the course of this technological innovation and as it increasingly diverges from a firm size that exists in reality but perhaps has become outdated under the protection of the cartel. The alliance now becomes an economic unit. The last consequence of the increase in an optimum position is fixed when the most favourable quantity supplied by the optimum firm either achieves or exceeds the total output (Translator's note: "total output" see Scherer: 65) of the branch of production concerned. Now in the long run, one enormous enterprise emerges which consists of a single enormous firm. This will approximately achieve its optimum at first. If it continues to grow, the enterprise eventually reaches cost degression to an increasing extent, or even the area of increasing returns.[22]

---

[22] The conclusions in this section supplement proposition (XVII). In conjunction with this proposition, they form the basis of an analysis of some theoretical ideas:

(a) Hence e.g. the diagrams in Alfred Marshall *Principles of Economics*, Figs. 24, 26, 28, 29, 30, 32, 33 and 35 appear to serve no purpose because they are based partly on assumptions about the competitive economy (in particular, they assume that the intersection point of the supply and demand curve produces the equilibrium point (Translator's note: "equilibrium point" see Heertje, A.: 48) but also partly on assumptions that are incompatible with the competitive economy (in particular, decreasing supply curves). However, the diagrams can only be rescued if we interpret the demand curve as being a marginal revenue curve rather than a price curve and interpret the supply curve to be a marginal cost curve and this phenomenon incidentally applies to the monopoly.

(b) Cassel, G. (1932): 86–88 (Translator's note: The exact quotation from the English translation "The Theory of Social Economy" by Cassel, G. (tr. McCabe, J. 1923: 102) has been used)) selects the following proposition as his second supplementary proposition of pricing. "When a larger output means cheaper production—when, that is to say, the average cost of the article, based upon the total production, is lowered by an increased production—the price of the article must, for the sake of equilibrium, correspond to the average cost of production". This statement equates to the principle of the satisfaction of needs and wants but is irreconcilable with the profit-making principle in free competition. Indeed, the price and average cost equation applies to the competitive enterprise, as shown above. These enterprises are however subject to the law of diminishing returns. The law of increasing returns is only possible in the profit-making economy with a monopoly. As we have shown, the price and average cost equation probably applies formally, not only because we must also see the price for entrepreneurial services as a component of cost in equilibrium, no matter whether this is a competitive price or a monopoly price.

Where technological progress in all branches of production leads to perpetual growth in the optimum firm size, the result will be to mass together the entire economic production setup into a single structure which only responds to one interest and as a result, can be described as one enterprise. The principle of the satisfaction of needs and wants applies within this enterprise, as was shown in the theory of transfer prices.

This entire market economy enterprise would imply aggregating all the elements of the relevant market economy because any remaining competitive branches of production would also proceed with monopolisation under the pressure of the general tendency to form a monopoly. This also applies to suppliers of the factors of production. This formally conceived entire enterprise would in practice simply be a tool of the state, engaging in equal measure in the market economy processes of production and allocation as and where they are concentrated. With a reversal in technological innovation, it would also be left to the state to reduce the size of the enterprise (it would be unlikely to increase the size of the enterprise) and to introduce the competitive organisation of the production system, which now becomes possible again.

References

Cassel, G. (1932). *Theoretsiche Sozialökonomie* (published in English as: *The Theory of Social Economy*): 86–88.

Clark, J.B. (1922). *Essentials of Economic Theory*: 132, The MacMillan Company, New York, p. XI+561.

Moore, H.L. (1929). *Synthetic Economics*. 145.

Schneider, E. (1932). *Reine Theorie monopolistischer Wirtschaftsformen*, (published in English as: *The Pure Theory of Monopoly*) Tübingen.

Walras, L. (1926). *Éléments d'économie politique pure* (published in English as: *Elements of Pure Economics*): 175 et seq.

Appendix A. Mathematical Appendix

In the following pages, the evidence presented in the text has been supplemented using analytical derivations.

I. The Optimum Position (Chap. 2, § 2)

1. Optimum velocity of production p is defined by the fact that its average costs are at a minimum. p results from the following equation:

$$\frac{d\,K^*}{d\,x}\{p\} = 0$$

or using $K^* = \frac{K}{x}$ from equation

$$\frac{K'\,(p)}{p} - \frac{K\,(p)}{p^2} = 0,$$

whereby after some remodelling, it follows that:

$$K'\,(p) = K^*\,(p). \tag{1}$$

This is the fundamental proposition which underlies the optimum position (proposition I).

2. By definition, the minimum condition for the optimum position follows:

$$\frac{d^2\,K^*}{d\,x^2}\{p\} > 0. \tag{2}$$

H. von Stackelberg, *Foundations of a Pure Cost Theory*,
DOI 10.1007/978-3-642-34537-1,

It is however:

$$\frac{d^2 K^*}{d x^2} = \frac{K''(x)}{x} - \frac{2\,[K'(x) - K^*(x)]}{x^2}.$$

Using (1), it therefore follows that:

$$\frac{d^2 K^*}{d x^2}\{p\} = \frac{K''(p)}{p}.$$

As a result, (2) is synonymous with

$$K''(p) > 0. \tag{3}$$

This inequality contains proposition II.

Where multiple values of x satisfy conditions (1) and (3), p is the value which gives the lowest value for function K^*.

3. According to the definition of K, K_{I} and K_{II}, it follows that:

$$K(x) = K_{\mathrm{I}} + \int_0^x K'(\xi)\, d\,\xi.$$

Using (1), we have:

$$p\,.\,K'(p) = K_{\mathrm{I}} + \int_0^p K'(\xi)\, d\,\xi$$

because when $x = b$, by definition $K(x)$ becomes minimised so that:

$$K'(x) \geqq K'(b)$$

that is to say,

$$\int_0^P K'(\xi)\, d\xi \geqq \int_0^p K'(b)\, d\xi = p\,.\,K'(b).$$

We therefore derive:

$$p\,.\,K'(p) \geqq K_{\mathrm{I}} + p\,.\,K'(b)$$

or

$$K'(p) - K'(b) \geqq \frac{K_I}{p}.$$

When $x = p$, function K' has already increased and it must therefore follow that $p > b$, and indeed, by an amount that requires an increase from $K'(b)$ to $K'(p)$ of at least $\frac{K_I}{p}$.

4. To prove proposition III, we analytically formulate the following assumptions:

α) $K''(x)$ exists and is continuous for $x \geq 0$ (regularity and smoothness)
β) $+K''(x) = +(x - b)$ (regularity).

The assertion then is:

$$\text{sign}\,[K'(x) - K^*(x)] = \text{sign}\,(x - p). \tag{4}$$

We firstly look at function

$$f(x) = x\,[K'(x) - K^*(x)] = x \,.\, K'(x) - K(x).$$

It follows that:

$$\text{sign}\, f(x) = \text{sign}\,\frac{f(x)}{x} = \text{sign}\,[K'(x) - K^*(x)].$$

Our assertion is then established once we have established this equation:

$$\text{sign}\, f(x) = \text{sign}\,(x - p).$$

It also follows that:

$$\lim_{x\to 0} K^*_{II}(x) = \lim_{x\to 0} \frac{K(x) - K(o)}{x} = K'(o). \tag{5}$$

As a result, we derive:

$$\lim_{x\to 0} f(x) = \lim_{x\to 0} x \,.\, \lim_{x\to 0} K'(x) - \lim_{x\to 0} K(x) = -K_I.$$

We note $f(o) = \lim_{x\to 0} f(x)$, as a result of which, $f(x)$ also becomes continuous for $x = o$ and it therefore follows that:

$$f(o) = -K_I < 0. \tag{6}$$

In addition:

$$\frac{df(x)}{dx} = x \, . \, K''(x). \tag{7}$$

Using condition β), it follows that $f(x)$ monotonically decreases for $x < b$ and monotonically increases for $x > b$ with a minimum position for $x = b$. Using (6), $f(x)$ is therefore negative for $x < b$. Using (1), $f(p) = 0$ and consequently using (7) it is positive for $x > p$ and negative for $b < x < p.f(b)$ is negative as a minimum for negative values too. It therefore follows that:

$$\operatorname{sign} f(x) = \operatorname{sign}(x - p).$$

As a result, we have established assertion (4).

5. Proposition III includes proposition IIIb. It follows from the simple observation that an enterprise with a constant increase in revenue is also subject to cost degression. As a condition, we can assume that $K'(x) =$ constant. Hence:

$$K(x) = K_{\mathrm{I}} + \int_0^x K' \, d\xi = K_{\mathrm{I}} + x \, . \, K'$$

$$K^*(x) = \frac{K_{\mathrm{I}}}{x} + K'$$

$$K'(x) - K^*(x) = -\frac{K_{\mathrm{I}}}{x} < 0, . \text{ Q.E.D.}$$

II. The Minimum Position (Chap. 2, § 3)

1. The conditional equation for q follows from similar explanations to those in (1) and using $K'(x) = K'_{\mathrm{II}}(x)$:

$$K'(q) = K^*_{\mathrm{II}}(q). \tag{8}$$

This equation expresses proposition IV.

2. Inequality (2) now equates to this inequality:

$$\frac{d^2 K^*_{\mathrm{II}}}{d\,x^2}\{q\} > 0. \tag{9}$$

As with inequality (3), the following inequality then results:

$$K''(q) > 0. \tag{10}$$

This is the assertion in proposition V.

3. Proposition VI is established using (5).

III. The Enterprise's Supply According to the Profit-making Principle (Chap. 2, § 4)

1. By definition, the most favourable velocity of production s maximises profit. It now follows that:

$$G'(s) = E'(s) - K'(s) = 0$$

or

$$E'(s) = K'(s). \tag{11}$$

This is the fundamental proposition of the profit-making principle.

2. The second maximum condition for G is:

$$G''(s) < 0. \tag{12}$$

In other words, $G'(x)$ decreases in the vicinity of $x = s$ or using (11), it is positive when $x < s$ and negative when $x > s$. Proposition XI therefore follows for a sufficiently small value of s.

3. Proposition XIII is established as follows:
According to its assumptions, the next inequality would follow for all x:

$$E'(x) - K'(x) = G'(x) < 0. \tag{13}$$

The profit function now decreases monotonically with increasing x. Its maximum therefore occurs when $x = 0$. It would be best for the enterprise if it were to lie idle. It would then incur the smallest loss, that is, K_{I}.

4. Using $K'_{\mathrm{II}} = K'$, (11) can be replaced with equation

$$E'(s) = K'_{\mathrm{II}}(s). \tag{14}$$

We therefore obtain proposition XV.

5. Under free competition, price P is independent of x. Then $E(x) = x.P$ and $E'(x) = P$. Now (11) is replaced by

$$P = K'(s). \tag{15}$$

and (15) is proposition XVI. (12) gives:

$$K''(s) > 0 \tag{16}$$

from which propositions XVII and XVIII follow.

6. In the case of a monopoly, the price of the goods produced and supplied by the enterprise is a monotonically decreasing function of the velocity of production. Therefore the following applies:

$$P'(x) < 0. \tag{17}$$

From (11) follows:

$$K'(s) = P(s) + s \,.\, P'(s). \tag{18}$$

If we describe the absolute amount $|P'(x)|$ for $P'(x)$ as the slope of the demand function, then using (17) and (18), it therefore follows that:

$$K'(s) = P(s) - s \,.\, |P'(s)|. \tag{19}$$

These are the contents of proposition XX.

We obtain a different formulation of this proposition when we use a common term in theoretical economics, the "elasticity of demand". We do not want to derive the analytical term for the measure of elasticity (Translator's note: "measure of elasticity" for a more in-depth discussion of this question, cf. Baumgärtner Stephan, (2001), "Heinrich von Stackelberg on Joint Production", The European Journal of the History of Economic Thought, 8(4): 509–525, winter.) ourselves, but will instead take it directly from Alfred Marshall's *Principles of Economics*.[1] Marshall expresses this term as follows: $\frac{dx}{x} : \frac{-dy}{y}$. Here, x is the quantity of goods which is therefore identical to our x while y is the price which is therefore identical to our P. We will term elasticity: $\frac{dx}{x} : \frac{-dP}{P}$. We will simply refer to it as "elasticity".[2] We will use the symbol ε to describe it. ε is a function[3] of x:

$$\varepsilon(x) = \frac{dx}{x} : -\frac{dP}{P} = -\frac{P(x)}{x \,.\, P'(x)}. \tag{20}$$

[1] Marshall, A. 685; see also: Dalton, H. 192 et seq.

[2] See also Bowley, A. ibid.: 32–33.

[3] By integrating this differential equation, we establish an important analytical expression for the price function in relation to so-called "synthetic economics". Its derivation is given below:

We will assume that $\lim\limits_{x \to 0} \varepsilon(x)$ exists and is other than zero. $\lim\limits_{x \to 0} \varepsilon(x) \to \infty$ would also be possible. Then let $\frac{1}{\varepsilon(x)} = a + \eta(x)$, which would be $a = \lim\limits_{x \to 0} \frac{1}{\varepsilon(x)}$. Then:

The following is revealed from (20):

$$x \,.\, P'(x) = -\frac{P(x)}{\varepsilon(x)}.$$

As a condition for the most favourable velocity of production, using (18), the following occurs:

$$P(s) - K'(s) = \frac{P(s)}{\varepsilon(s)}. \tag{21}$$

These are the contents of proposition XXa. Proposition XXI arises directly from (19) or (21).

IV. The Costs of Joint Production (Chap. 3, §§ 1 to 3)

1. Above all, it is a question of finding the coordinates system which corresponds best to our subject. If we regard velocities of production as being Cartesian coordinates, then polar coordinates are our initial choice for a system of coordinates.

   Therefore we regard cost functions in terms of their dependency on the variables r and φ, where $r = \sqrt{x_1^2 + x_2^2}$ and we have $\tan \varphi = \frac{x_2}{x_1}$. If (x_1, x_2) is regarded as vector $\mathfrak{x}$, r is then its length and φ is its direction. We will therefore refer to "length of production" r and "direction of production" φ. Let $K(x_1, x_2) = K[r, \varphi]$ and correspondingly, let $P(x_1, x_2) = P[r, \varphi]$. We describe the vector $(\cos \varphi, \sin \varphi)$ as vector component $\mathfrak{e}$. It is then $\mathfrak{x} = \mathfrak{e}.\, r$. In particular, we now have:

$$\frac{P'}{P} = -\frac{1}{x \,.\, \varepsilon(x)} = -\frac{a}{x} - \frac{\eta(x)}{x}$$

$$\int_0^x \frac{P'}{P}\, d\xi = -a \int_0^x \frac{d\xi}{\xi} - \int_0^x \frac{\eta(\xi)}{\xi}\, d\xi + \ln C$$

$$\ln P = a \ln x + \ln C - \int_0^x \frac{\eta(\xi)}{\xi}\, d\xi$$

$$P = \frac{C}{x^a} \cdot e^{-\int_0^x \frac{\eta(\xi)}{\xi} d\xi}$$

Where $\varepsilon(x)$ is constant, the factor $e^{-\int_0^x \frac{\eta(\xi)}{\xi} d\xi}$ has the value 1. Then $\varepsilon(x) = \frac{1}{a}$ for all values of x. In this case, we must note that the revenue function $x.\, P$ only starts from the origin according to our assumption if $a < 1$ and therefore if $\varepsilon > 1$, that is, if demand is elastic.

(a) the total cost function:

$$K\,[r,\ \varphi] = K\,(r\,.\,\cos\,\varphi,\ \ r\,.\,\sin\,\varphi);$$

(b) the marginal cost function:

$$\frac{\partial K}{\partial r} = K'_r[r,\ \phi] = K'_1\,.\,\cos\ \phi + K'_2\,.\,\sin\ \phi,$$
$$\text{where } K'_1 = \frac{\partial K}{\partial x_1} \text{ and } K'_2 = \frac{\partial K}{\partial x_2};$$

correspondingly, the marginal cost gradient;
(c) average costs:

$$K^*\,[r,\ \varphi] = \frac{K}{r};$$

correspondingly, average variable costs;
(d) returns in the general case of a monopoly:

$$\begin{aligned} E[r,\varphi] &= E\,(r\,.\cos\varphi,\, r\,.\sin\varphi) \\ &= r\,.\cos\varphi\,.\,P_1\,(r\,.\cos\varphi,\, r\,.\sin\varphi) + r\,.\sin\varphi\,.\,P_2\,(r\,.\cos\varphi,\, r\,.\sin\varphi) \\ &= r\{\cos\varphi\,.\,P_1\,[r,\varphi] + \sin\varphi\,.\,P_2\,[r,\varphi]\} \\ &= \mathfrak{r}\,.\,\mathfrak{P}. \end{aligned} \qquad (22)$$

Special cases result from this formula.

2. Where the composition ratio x_1: x_2 is constant for joint production, we then have to regard vector component $\mathfrak{e}$ for the (constant) direction of production as a component of the quantity of product. Then r is the quantity of units of product (Translator's note: "units of product" see Cassel: 280) produced in the unit time, that is, the rate of production in the case of single production. As a result, the fundamental proposition of joint production (XXVI) follows.
3. A length of production b, q, p and s is assigned respectively to each direction. We hence obtain b, q, p and s as functions of direction φ. It follows in particular that:
   if we let

$$\frac{\partial^2 K}{\partial r^2}\{r,\ \varphi\} = 0,$$

then $r = b$. We have therefore defined variable b as an implicit function of φ using equation

$$\frac{\partial^2 K}{\partial r^2}\{b,\ \varphi\} = 0.$$

$r = b\ [\varphi]$ is the equation of the b curve. Likewise

$$\frac{\partial K}{\partial r}\{q,\ \varphi\} - K^*_{\mathrm{II}}\ [q,\ \varphi] = 0$$

is the definition of $r = q\ [\varphi]$ (q curve) and

$$\frac{\partial K}{\partial r}\{p,\ \varphi\} - K^*\ [p,\ \varphi] = 0$$

is the definition of $r = p\ [\varphi]$ (p curve) as well as (eventually)

$$\frac{\partial E}{\partial r}\{s,\ \varphi\} - \frac{\partial K}{\partial r}\{s,\ \varphi\} = 0$$

is the definition of $r = s\ [\varphi]$ (s curve) and consequently, the most favourable velocities of production curve.

If we regard direction as being along concentric circles, that is, along curves with the equation $r =$ constant, we obtain the most favourable direction curve from the following equation:

$$\frac{\partial E}{\partial \varphi}\{r,\ \sigma\} - \frac{\partial K}{\partial \varphi}\{r,\ \sigma\} = 0.$$

From this we obtain $\varphi = \sigma\ [r]$. The point which satisfies the two equations $r = s\ [\varphi]$ and $\varphi = \sigma\ [r]$, represents the most favourable production level. We therefore have the three propositions XXVII to XXIX.

4. The polar coordinates system consists of a family of concentric circles around the origin and a family of radii originating at the origin. We replace the family of circles with the family of indifference cost curves. We then obtain a coordinates system that corresponds to our subject as best it can. A product vector is now determined using the level of its production costs and its direction of production, that is, using the proportion of its components. We eventually bring the family of indifference cost curves face to face with the family of indifference revenue curves.

   The family of indifference cost curves has the equation:

$$K\,[r,\ \varphi] = \mathrm{M} \tag{23}$$

where M is the parameter of the family. The family of indifference revenue curves has the equation:

$$E[r,\ \varphi] = L \tag{24}$$

where L is the parameter of this family.

An arbitrary indifference cost curve k has the equation $K\,[r,\varphi]-\mathrm{M}_k = 0$ and defines r as being a definite function of φ, therefore:

$$r = k\,[\varphi]. \tag{25}$$

By inserting (25) into return function (22), we arrive at curve k's return as a function of direction:

$$E = E\,[k\,[\varphi],\ \varphi] = E\ [\varphi].$$

Its maximum follows in this way:

$$\frac{dE}{d\varphi} = \frac{\partial E}{\partial r}\cdot\frac{dk\,[\varphi]}{d\varphi} + \frac{\partial E}{\partial \varphi} = \frac{\partial E}{\partial r}\cdot\left(-\frac{\frac{\partial K}{\partial \varphi}}{\frac{\partial K}{\partial r}}\right) + \frac{\partial E}{\partial \varphi} = 0$$

$$\frac{\partial E}{\partial r}\cdot\frac{\partial K}{\partial \varphi} - \frac{\partial E}{\partial \varphi}\cdot\frac{\partial K}{\partial r} = 0$$

or

$$\begin{pmatrix} \frac{\partial E}{\partial r}, & \frac{\partial E}{\partial \varphi} \\ \frac{\partial K}{\partial r}, & \frac{\partial K}{\partial \varphi} \end{pmatrix} = 0. \tag{26}$$

This term applies to each indifference cost curve. It is the equation of the curve which is made up of the most favourable points on the indifference cost curves. This equation states that disregarding any points occurring at the edge of the area defined by the total cost function and the revenue function, the most favourable point of each indifference cost curve is characterised by the fact that the indifference cost curve and the indifference revenue curve which pass through this point also have a common tangent at this point.

Equation (26) can be generalised in the case where n goods are produced jointly. We then derive the length of production r and $n-1$ angles $\varphi_1, \varphi_2 \dots \varphi_{n-1}$ as coordinates which together equate to the direction of production. Equation (26) states that the two-dimensional matrix

$$\begin{pmatrix} \frac{\partial E}{\partial r}, & \frac{\partial E}{\partial \varphi} \\ \frac{\partial K}{\partial r}, & \frac{\partial K}{\partial \varphi} \end{pmatrix}$$

has the rank 1. The same applies to the case of n goods for the corresponding n-dimensional two-row matrix:

$$\begin{pmatrix} \frac{\partial E}{\partial r}, & \frac{\partial E}{\partial \varphi_1}, & \dots & \frac{\partial E}{\partial \varphi_{n-1}} \\ \frac{\partial K}{\partial r}, & \frac{\partial K}{\partial \varphi_1}, & \dots & \frac{\partial K}{\partial \varphi_{n-1}} \end{pmatrix}.$$

The two-row determinants of this matrix are set at zero where we obtain $n-1$ equations which together define a spatial curve in n-dimensional space.[4]

V. Theory of the Inter-Firm Transfer Price (Chap. 3, § 4)

We use the same terms that appear in the text. We use H to represent the costs from firm 2 that were added to the costs of firm 1. Clearly, H depends on x_2 and y. The variable y is however defined as a function of x_1 and x_2 using the condition that the costs of x_1-y and x_2 must be at a minimum. We therefore derive the following for the entire enterprise:

$$K = K_1(x_1) + H(x_2, y) \tag{27}$$

and for firm 2:

$$K_2 = y \,.\, V + H(x_2, y) \tag{28}$$

where simultaneously this condition[5] follows:

$$\frac{\partial H}{\partial y} = -\frac{dK_1}{dx_1}. \tag{29}$$

Using (29), y is defined as a function of x_1 and x_2.

1. Where good 1 is not marketable, then $y = x_1$. In order that the entire enterprise maximises profits, it must satisfy this condition:

---

[4] Equation (26) can also be written as $\frac{\partial K}{\partial r} : \frac{\partial K}{\partial \varphi} = \frac{\partial E}{\partial r} : \frac{\partial E}{\partial \varphi}$. It corresponds exactly to Edgeworth's "contract curve" equation that was cited on pp. 690 in the German edition of Marshall's *Principle of Economics*; it has a different but similar meaning here. A simple calculation incidentally shows that determinant (26) has the same value as determinant $\begin{vmatrix} E'_1 & E'_2 \\ K'_1 & K'_2 \end{vmatrix}$. It is hence invariant with regard to our coordinate transformation.

[5] This condition occurs as follows: if we let $x_1 - y = u$, then for a particular u and x_2, the variable K will become a minimum. Consequently:

$$\frac{\partial K(u, y, x_2)}{\partial y} = \frac{\partial K_1(u+y)}{\partial y} + \frac{\partial H(x_2, y)}{\partial y} = 0$$

from which we obtain formula (29) using $\frac{\partial K_1(u+y)}{\partial y} = \frac{dK_1(x_1)}{dx_1}$

$$\frac{dE}{dx_2}\{s_2\} = \left[\frac{dK_1}{dy} + \frac{\partial H}{\partial y}\right] \cdot \frac{dy}{dx_2}\{s_2\} + \frac{\partial H}{\partial x_2}\{s_2\}. \tag{30}$$

Assuming that the profit-making principle is a prerequisite for firm 2, it will orientate itself to the market using its production.[6] Using (28), for firm 2 it follows that:

$$\frac{dE}{dx_2}\{s_2\} = \left[\frac{d(y\,.\,V)}{dy} + \frac{\partial H}{\partial y}\right] \cdot \frac{dy}{dx_2}\{s_2\} + \frac{\partial H}{\partial x_2}\{s_2\}. \tag{31}$$

The following equation results from (30) and (31) for each point s_2:

$$\frac{dK_1(y)}{dy} = \frac{d(y\,.\,V)}{dy}. \tag{32}$$

If (32) is to be fulfilled for each alternative value s_2, it must generally follow that: $y\,.\,\mathrm{V} = \mathrm{K}_1(y) + \mathrm{L}$, where L is an arbitrary constant. In line with proposition XXXV, we obtain the formula:

$$V = K_1(y) + \frac{\mathrm{L}}{y}. \tag{33}$$

2. Good 1 would be marketable. The transfer price will be determined in such a way that firm 2 always finds itself in agreement with the interests of the entire enterprise if it supplies in accordance with the profit-making principle. We must abandon setting the supply for firm 1 and make this directly dependent on the profit function of the entire enterprise.

   If prices P_1 and P_2 depend on the market supply of both goods, then:

$$P_1 = P_1(x_1 - y,\ x_2) \text{ and } P_2 = P_2(x_1 - y,\ x_2). \tag{34}$$

It then generally follows that:

$$G = G(x_1,\ y,\ x_2) \text{ and } G_2 = G_2(x_1,\ y,\ x_2).$$

Therefore y is defined using minimum condition (32) for the production costs of x_2 as a function of x_2.

The maximum condition for G is:

$$\frac{\partial G}{\partial y} \cdot \frac{\partial y}{\partial x_1} + \frac{\partial G}{\partial x_1} = 0; \quad \frac{\partial G}{\partial y} \cdot \frac{\partial y}{\partial x_2} + \frac{\partial G}{\partial x_2} = 0. \tag{35}$$

[6] See Chap. 3, § 4, I.

(a) For firm 1, y is a given variable which is ordered by firm 2. x_2 is also a given for this firm. Firm 1 then only directly influences the profit of the entire enterprise using total production s_1 of its product. We obtain s_1 from the first equation (35) which, after some remodelling and taking (29) into account, gives:

$$(s_1 - y) \cdot \frac{\partial P_1}{\partial x_1} + P_1 + x_2 \cdot \frac{\partial P_2}{\partial x_1} - K_1' \, (s_1) = 0. \tag{36}$$

Equation (36) defines s_1 as an implicit function of y and x_2.

(b) Firm 2 makes an offer on the assumption that the profit-making principle is a prerequisite. Using (32) and (36), its profit only depends on x_2 in the end. The maximum condition for G_2 as a conditional equation for s_2 is:

$$\frac{\partial G_2}{\partial x_1} \cdot \frac{d s_1}{d x_2} + \frac{\partial G_2}{\partial y} \cdot \frac{d y}{d x_2} + \frac{\partial G_2}{\partial x_2} = 0. \tag{37}$$

At point $x_1 = s_1$ and $x_2 = s_2$, G and G_2 must be at a maximum. Equations (35) and (37) must then be fulfilled for the same value x_1 and x_2 if firm 2's reaction is to maximise profits for the entire enterprise. This is the case for all prices and price functions. This requirement is fulfilled by equating the derivatives of G_2 to the corresponding derivatives of G. We then obtain these three identities:

$$\frac{\partial G_2}{\partial x_1} = \frac{\partial G}{\partial x_1}; \; \frac{\partial G_2}{\partial y} = \frac{\partial G}{\partial y}; \; \frac{\partial G_2}{\partial x_2} = \frac{\partial G}{\partial x_2}. \tag{38}$$

The assertion of proposition XXXVI follows from (38) if L is an arbitrary constant:

$$G - G_2 = \mathrm{L}. \tag{39}$$

Noting (27), (28) and (34), we obtain the following:

$$G = (x_1 - y) \cdot P_1 + x_2 \cdot P_2 - K_1 - H$$
$$G_2 = x_2 \cdot P_2 - y \cdot V - H.$$

The following results from this and from (39):

$$V = P_1 - \frac{x_1 \cdot P_1 - K_1}{y} + \frac{\mathrm{L}}{y}. \tag{40}$$

3. Equation (40) is the general formula for the inter-firm transfer price.[7] This can be derived from (40) for all special cases.

   (a) If good 1 is not marketable, its price P_1 is not initially defined. If we let $P_1 = 0$, we then have

$$V = \frac{K_1}{y} + \frac{\mathrm{L}}{y}$$

   from which we obtain formula (33) using $x_1 = y$.

   (b) If free competition exists in the market for good 1, then (36) becomes $P_1 = K'\ (s_1)$, from which s_1 follows independently from y and x_2. From (40), we then obtain:

$$V = P_1 - \frac{s_1\ .\ P_1 - K_1 - \mathrm{L}}{y}.$$

   The numerator in the fraction on the right side is independent of y and x_2. We can denote it using -M where M is a constant. Then proposition XXXVII follows:

$$V = P_1 + \frac{\mathrm{M}}{y}. \tag{41}$$

   (c) If P_1 is independent of x_2, and P_2 is independent of x_1, then using (36) we obtain:

$$(s_1 - y)\ .\ \frac{\partial\ P_1}{\partial\ x_1}\{s_1 - y\} + P_1\ (s_1 - y) - K_1'\ (s_1) = 0. \tag{42}$$

   (42) defines s_1 as an implicit function of y. The formula for the transfer price results from (40):

[7] It can easily be shown that condition (29) is satisfied for firm 2 using the transfer price to determine the current lowest costs and if $x_1 = u + y$, it follows that:

$$\frac{\partial K_2\,(u,\ y,\ x_2)}{\partial y} = \frac{\partial\,(y\ .\ V)}{\partial y} + \frac{\partial H}{\partial y}.$$

Using (4)

$$y\ .\ V = -u\ .\ P\,(u) + K_1\,(u + y),$$

then

$$\frac{\partial\ (y\ .\ V)}{\partial y} = \frac{\partial K_1}{\partial y} = \frac{\partial K_1}{\partial x_1}.$$

As a result, we arrive at (29).

$$V = P_1\ (s_1 - y) - \frac{s_1\ .\ P_1\ (s_1 - y) - K_1\ (s_1)}{y} + \frac{\mathrm{L}}{y}. \tag{43}$$

Here V is fundamentally different to market price P_1, that is, it is not only different regarding the quotients of a constant and y, because the numerator in the first fraction in (43) by way of $P_1\ (s_1-y)$ and s_1 depends on y using (42). The observations in § 4, IV that lead to proposition XXXVIII follow from this. Incidentally, using (42), it is established that firm 1's supply is independent of firms 2's supply in the sense that firm 1 does not need to orientate itself to firm 2's market situation for its market supply, but instead can make calculations independently. The cost calculation can now be broken down.

(d) Where the price and the maximum saleable quantity for good 1 are fixed (e.g., through a cartel) and we describe the maximum saleable quota using h, we can then distinguish two cases:

(α) If firm 1's supply s_1 is greater than $h + y$, which would occur if sales (at given prices) were free and independent, firm 1 will produce quantity $h + y$. Using (40) as well as $x_1 = h + y$, we then derive

$$V = \frac{K_1\ (h+y)}{y} + \frac{\mathrm{L} - h\ .\ P_1}{y}. \tag{44}$$

Since $\mathrm{L}-h.\ P_1$ is constant, (44) therefore has a certain similarity to (33).

(β) Where $s \leq h + y$, then the same quantity is produced as under free competition, that is, according to formula (41).

(e) A specific situation occurs in the case of modified rivalry. Here we can regard total costs as being the sum of two functions: $K_1\ (x_1) + C\ (x_1-y)$, where C represents the sales costs which are not possible for y; they depend on the quantity sold $x_1 - y$. Now (36) appears in this form:

$$P_1 - K_1'\ (s) - C'\ (s_1 - y) = 0.$$

As a result, s_1 is defined as a function of y. From (40) follows:

$$V = P_1 - \frac{s_1\ .\ P_1 - K_1 - C\ (s_1 - y)}{y} + \frac{\mathrm{L}}{y}.$$

The numerator from the first fraction is dependent on y by virtue of C and s_1. We then have a similar case to the one in (43), that is, the transfer price is fundamentally different to the market price.

4. The problem becomes increasingly complex if both firms produce multiple goods. Now we must apply vector calculus to provide a clearer description. We describe the quantity produced by firm 1 using x_k (k = 1, 2, ..., n), the quantity ordered by firm 2 using y_k (k = 1, 2, ... n), the quantity produced by firm 2 (all in the unit time) using z_i (i = 1, 2, 3 ....., n), and furthermore vector $(x_1, x_2, ... x_n)$ using $\mathfrak{x}$, vector $(y_1, y_2, ..., y_n)$ using $\mathfrak{y}$, vector $(z_1, z_2, ... z_m)$ using $\mathfrak{z}$, the price vector of all $n + m$ goods using $\mathfrak{P}$ and ultimately transfer prices using V_k (k = 1, 2, ... n), and the vector $(V_1, V_2, ... V_n)$ using $\mathfrak{P}$, so where we assign a coordinate in $(n + m)$-dimensional spaces for each good, it follows that:

$$G = (\mathfrak{x} - \mathfrak{y} + \mathfrak{z}) \ . \ \mathfrak{P} - K_1 \ (\mathfrak{x}) - H \ (\mathfrak{y}, \ \mathfrak{z}) = G \ (\mathfrak{x}, \ \mathfrak{y}, \ \mathfrak{z})$$
$$G_2 = \mathfrak{z} \ . \ \mathfrak{P} - \mathfrak{y} \ . \ \mathfrak{P} - H \ (\mathfrak{y}, \ \mathfrak{z}) = G_2 \ (\mathfrak{x}, \ \mathfrak{y}, \ \mathfrak{z}).$$

It must be $G_2 = \{0\}$ degrees when and only when $G = \{0\}$ degrees. This is achieved by letting G_2 degrees $= G$ degrees. Then G_2 can only be distinguished from G by an arbitrary constant L. We therefore derive:

$$\mathrm{L} = (\mathfrak{x} - \mathfrak{y}) \ . \ \mathfrak{P} - K_1 \ (\mathfrak{x}) + \mathfrak{y} \ . \ \mathfrak{B}$$

and ultimately:

$$\mathfrak{y} \, . \, \mathfrak{B} = (\mathfrak{y} - \mathfrak{x}) \ . \ \mathfrak{P} + K \ (\mathfrak{x}) + \mathrm{L}.$$

In other words, firm 2 is burdened by firm 1 with the total costs of firm 1's production for all the quantities of goods ordered from firm 2 in the unit time for firm 1, and is reflected in firm 1's revenue in the market. An arbitrary constant is added to one of these items. A transfer price cannot generally be constructed for each individual good. One exception to this arises here if free competition predominates in firm 1's market and the other prices are also independent of x, that is, if this equation follows:

$$\sum_1^n {}_k \, y_k \ . \ V_k = - \sum_1^n {}_k \, (x_k - y_k) \ . \ P_k + K_1 \ (\mathfrak{x}) + \mathrm{L}.$$

Now the most favourable $\mathfrak{x}$ vector results from these equations:

$$G = \{0\} \text{ degrees}, \qquad \text{that is, from}: \ P_k = \frac{\partial \, K_1}{\partial \, x_k}.$$

We describe the roots of these n equations using s_k and vector $(s_1, s_2, ..., s_n)$ using $\mathfrak{s}$. s_k follow independently from $\mathfrak{y}$ and $\mathfrak{z}$. We now have:

$$\sum_{1}^{n}{}_{k}\, y_k \,.\, V_k = \sum_{1}^{n}{}_{k}\, y_k \,.\, P_k - \left[\sum_{1}^{n}{}_{k}\, s_k \,.\, P_k - K_1 \;(\mathfrak{s})\right] + \mathrm{L} \quad \text{or}$$

$$\sum_{1}^{n}{}_{k}\, y_k \,.\, V_k = \sum_{1}^{n}{}_{k}\, y_k \,.\, P_k \tag{45}$$

if we firstly let the arbitrary constant L be equal to the constant variable $\sum_{1}^{n}{}_{k}\, s_k . P_k - K_1$ ($\mathfrak{s}$). Equation (45) follows but only for each value of each y_k if $V_k = P_k$, that is, if:

$$\mathfrak{V} = \mathfrak{P}.$$

Under free competition, we then have the same situation for joint production as we did for single production. Nothing has changed if we let

$$V_k = P_k + \frac{\mathrm{M}_k}{y_k}.$$

If we describe the constant

$$\sum_{1}^{n}{}_{k}\, \mathrm{M}_k \quad \text{using} \quad \mathrm{M},$$

it therefore follows that:

$$\sum_{1}^{n}{}_{k}\, y_k \,.\, V_k = \sum_{1}^{n}{}_{k}\, y_k \,.\, P_k + \mathrm{M}$$

or:

$$\mathfrak{y} \,.\, \mathfrak{V} = \mathfrak{y} \,.\, \mathfrak{P} + \mathrm{M}.$$

VI. The Static Equilibrium with the Given Distribution of Enterprises (Chap. 4, § 1, I)

We assign an n-dimensional number range to the n types of goods in our system. Furthermore, we number our m economic entities and these are partly enterprises and partly ordinary individuals. We regard the supply of an economic agent μ to be a vector and denote it using $\mathfrak{x}_\mu$. The components of this vector are equal to zero where the corresponding goods are not supplied by the individual concerned. An enterprise's supply vector is just its product vector. Accordingly, we define vector $\mathfrak{y}_\mu$ to be the demand vector of economic agent μ. For an enterprise, $\mathfrak{y}_\mu$ is simply its input vector, where the capital used and the velocity of input are regarded as the availability of capital in the unit time. $\mathfrak{P}$ is the general price vector.

1. The Individual Equilibrium of the Enterprise[8]

The enterprise is assigned a production function which links the input vector and the product vector to each other. Since the availability of capital is contained in the input vector, the production function (Translator's note: "production function" see Baumgärtner: 518) depends on the price vector. It then has the general form[9]:

$$\varphi\ (\mathfrak{x} - \mathfrak{y}, \mathfrak{P}) = 0. \tag{46}$$

This function has the following meaning: because this is a question of free competition, $\mathfrak{P}$ is regarded as a given for the enterprise concerned. If any $\mathfrak{x}$ are produced, all $\mathfrak{y}$ which satisfy equation (46) with that $\mathfrak{x}$ are input vectors which are technically suited to producing $\mathfrak{x}$. Where an input vector $\mathfrak{y}$ is fixed, all $\mathfrak{x}$ which satisfy equation (46) with that $\mathfrak{y}$ are product vectors which can be produced with $\mathfrak{y}$.

The enterprise attempts to maximise the internal vector product.

$$(\mathfrak{x} - \mathfrak{y})\ .\ \mathfrak{P}$$

using constraint (46) where $\mathfrak{P}$ is constant, as was mentioned before. The maximum condition is now:

Matrix

$$\begin{pmatrix} \mathfrak{P} \\ \text{degrees}_{(\mathfrak{x}-\mathfrak{y})}\ \varphi \end{pmatrix} \tag{47}$$

is rank 1 for the realised values $\mathfrak{x}$ and $\mathfrak{y}$.

We can usually assume that components of the product vector which are other than zero, are zero components for the input vector and vice versa. The following equation applies to every partial derivative of a component y_k that exists for $\mathfrak{y}$ because of (46) according to a component x_i for $\mathfrak{x}$ ($i = 1, 2, \ldots.., n$; $k = 1, 2, \ldots.., n$):

$$\frac{P_i}{P_k} = \frac{\partial y_k}{\partial x_i} \text{ or } P_i = \frac{\partial y_k}{\partial x_i}\ .\ P_k. \tag{48}$$

[8] See Pareto, V. (1927), Appendix, No. 77 et seq.

[9] Here, we can omit the index which sets the number of the enterprise.

There are always precisely $n-1$ mutually independent (48) equations so that the vectors to be achieved by the enterprise with fixed price proportions are usually determined from (46) and (48). However, they are only determined in absolute terms for the capital component, apart from one factor which is dependent on the price level.

The following system of equations is synonymous with (48):

$$\frac{P_k}{P_i} = \frac{\partial x_i}{\partial y_k} \text{ or } P_k = \frac{\partial x_i}{\partial y_k} \,.\, P_i. \tag{49}$$

$\frac{\partial y_k}{\partial x_i}$ is the measure of quantity and $\frac{\partial y_k}{\partial x_i} \,.\, P_k$ is the measure of the value of the marginal input (Translator's note: "marginal input" see Peacock: 108) which must be achieved to produce product i using factor of production k. $\frac{\partial x_i}{\partial y_k}$ is the measure of quantity and $\frac{\partial x_i}{\partial y_k} \,.\, P_i$ is the value of the marginal return that is achieved using input of factor of production k to produce i.

Equation (48) therefore means that the value of the marginal input for a product produced from any factor of production is equal to the price of the product. Equation (49) similarly means that the value of the marginal return of a factor of production for every product is equal to the price of the factor of production. It follows from (48) that the values of the marginal inputs of all factors of production for each product are equal to each other. It follows from (49) that the value of marginal returns for all products are equal to each other for every factor of production.

2. *The Individual Equilibrium of the Ordinary Individual*

Ordinary individuals, who we shall describe as consumers, to distinguish them from the producing enterprise, are providers of factors of production and demanders of products. We can therefore distinguish supply vector $\mathfrak{x}$ and demand vector $\mathfrak{y}$ for this individual too. The individual's "equilibrium equation"[10] is:

$$(\mathfrak{x} - \mathfrak{y}) \,.\, \mathfrak{P} = 0. \tag{50}$$

An individual attempts to maximise his ordinal utility function or ordinal utility index function[11]

$$J = f(\mathfrak{x} - \mathfrak{y}, \mathfrak{P})$$

[10] See Pareto, V. (1927), Appendix, No. 80, equation (B).

[11] See Pareto, V. (1927), Appendix, No. 1 et seq.

using constraint (50), where $\mathfrak{P}$ is constant because we have assumed there is a competitive economy. The ordinal utility function applies to the consumer's overall situation, that is, it applies to his supply as well as his demand. Parameter $\mathfrak{P}$ is introduced because the capitalist's supply cannot be defined independently of the price vector.

We obtain a maximum condition similar to term (47):

The matrix

$$\begin{pmatrix} \mathfrak{P} \\ \text{degrees}_{(\mathfrak{x}-\mathfrak{y})} f \end{pmatrix}. \tag{51}$$

is rank 1 for the realised values $\mathfrak{x}$ and $\mathfrak{y}$.

In a similar way that (48) and (49) were derived from (47), the equations derived from (51) are as follows:

$$\frac{f'_{\mathrm{i}}}{P_{\mathrm{i}}} = \frac{f'_{\mathrm{k}}}{P_{\mathrm{k}}}. \tag{52}$$

These equations include the proposition established for equalising the level of marginal utility.[12]

3. *The General Equilibrium*

Aggregate supply vector $\mathfrak{A}$ for our n economic individuals is obtained using this term:

$$\mathfrak{A} = \sum_{1}^{\mathrm{m}} \mu\, \mathfrak{x}_\mu. \tag{53}$$

Aggregate demand vector $\mathfrak{N}$ is obtained using this term:

$$\mathfrak{N} = \sum_{1}^{\mathrm{m}} \mu\, \mathfrak{y}_\mu. \tag{54}$$

Each $\mathfrak{x}_\mu$ and each $\mathfrak{y}_\mu$ is defined as being a function of $\mathfrak{P}$ using (46) and (47) along with (50) and (51). Using (53) and (54), the same also follows for $\mathfrak{A}$ and $\mathfrak{N}$. Since the competition mechanism aims to equalise supply and demand, we then obtain the following term for price vector $\mathfrak{P}$ as a system of n conditional equations:

[12] See Pareto, A. (1927), Appendix, No. 80, equation (A).

$$\mathfrak{A}(\mathfrak{P}) = \mathfrak{N}(\mathfrak{P}). \tag{55}$$

This term determines $\mathfrak{P}$ except for a proportionality factor because, as we have already mentioned, $\mathfrak{x}$ and $\mathfrak{y}$ are independent of the absolute pricing level.

VII. The Proposition of Full Allocation in the Static Equilibrium (Chap. 4, § 1, III)

We take a simplified case that contains all the following main assumptions. Our enterprise produces a product using velocity of production x and price P, and three factors of production that have velocities of input y_1, y_2 and y_3 and prices Q_1, Q_2 and Q_3. All prices for the enterprise are fixed according to the assumption of competition. Then we can write the general production function (46) in a simplified form:

$$x = f(y_1, y_2, y_3). \tag{56}$$

Total costs for x are:

$$K(x) = y_1 . Q_1 + y_2 . Q_2 + y_3 . Q_3. \tag{57}$$

As usual, we describe the velocity of production achieved in reality using s and the velocities of input achieved using t_1, t_2 and t_3.

Profits are:

$$x . P - K(x) = P . f(y_1, y_2, y_3) - (y_1 . Q_1 + y_2 . Q_2 + y_3 . Q_3).$$

Maximum profits determine t_1, t_2 and t_3 and determine variable s by way of (56). The conditional equations for t_1, t_2 and t_3 are:

$$\left.\begin{aligned} P . f'_1(t_1, t_2, t_3) &= Q_1 \\ P . f'_2(t_1, t_2, t_3) &= Q_2 \\ P . f'_3(t_1, t_2, t_3) &= Q_3 \end{aligned}\right\}. \tag{58}$$

(58) is simply equation system (49). Proposition XXXIX states that prices in the static economy lead each enterprise to achieve its optimum position, that is, the minimum position of

$$\frac{K(x)}{x} = \frac{y_1 . Q_1 + y_2 . Q_2 + y_3 . Q_3}{f(y_1, y_2, y_3)}. \tag{59}$$

We obtain the minimum position for (59) from these three equations:

$$\left.\begin{aligned} s\,.\,Q_1 - f'_1(t_1,\, t_2,\, t_3)\,.\,K\,(s) = 0 \\ s\,.\,Q_2 - f'_2(t_1,\, t_2,\, t_3)\,.\,K\,(s) = 0 \\ s\,.\,Q_3 - f'_3(t_1,\, t_2,\, t_3)\,.\,K\,(s) = 0 \end{aligned}\right\}. \tag{60}$$

If we multiply the first of equations (60) by t_1, the second by t_2 and the third by t_3 and we then total them, we obtain:

$$s\,.\,K\,(s) - (t_1\,.f'_1 + t_2\,.f'_2 + t_3\,.f'_3)\,.\,K\,(s) = 0$$

or ultimately after remodelling:

$$f\,(t_1,\, t_2,\, t_3) = t_1\,.f'_1 + t_2\,.f'_2 + t_3\,.f'_3. \tag{61}$$

This means that in the static equilibrium state, the total of the quantity of factors of production used in the unit time is equal to their marginal productivity multiplied by total production. The equations in (58) show that in a competitive economy, allocations are made according to marginal productivity. Equation (61) shows that this allocation works out completely.

Equation (61) is a necessary but insufficient condition for a static competitive general equilibrium to exist. It follows that where the general production function of a market economy is obtained in such a way that (61) is not fulfilled for any input vector, the competitive organisational structure cannot be achieved in conjunction with the profit-making principle. Here, we can very clearly see the dependency of the socio-economic organisational structure on production functions, that is, on the technological situation.

Appendix B. Generalisation of the Total Cost Function[13]

Most of the propositions we have derived do not apply to every possible total cost function. We have specifically mentioned the condition of regularity. The propositions we have derived are therefore firstly only valid for total cost functions which satisfy this condition. We shall now investigate the extent to which those propositions can also be extended to the general case.

In giving the reasons for proposition XXII, we mentioned as an aside that very small values in economic terms can be ignored. We now turn this into a fundamental principle:

Any arbitrary adjustment to the total cost function is allowed, providing the profit function is only adjusted by a negligible amount α. Therefore, α can be determined as being insignificantly small (e.g. $\alpha = 0.0001$ Pfennig or pennies).

It is obvious that we are able to determine this and that the economic results remain unaffected by it.

Now we can prove the next proposition:

(XXXX). The total cost function can always be described using an unambiguous, monotonically increasing function which only has a finite number of points of discontinuity within every arbitrary finite interval and where it is continuous, it is also regular and smooth.

The total cost function is unambiguous and monotonical[14] and it already follows that discontinuities only occur due to a jump. The total cost function has two marginal values (Translator's note: "marginal values" see Peacock: 40) at the points of discontinuity, one on the left and one on the right. The one on the left is always smaller. We shall establish that the marginal value on the left is also always the function value of the point of discontinuity (which is always possible because of range α). The difference between the two marginal values is the jump, or as it is

[13] We now consider the theory of the length of production. We can make a very similar generalisation about the theory of the direction of production as before.

[14] Chap. 1, § 1, III, 2.

DOI 10.1007/978-3-642-34537-1,

generally referred to, the variation. A point of discontinuity is characterised by the fact that its variation is other than zero.

Where we have two arbitrary velocities of production x_1 and x_2, using the monotonicity of the total cost function, it follows that the total variations for all points of discontinuity in the interval (x_1, x_2) cannot be greater than the difference between the two total cost values (Translator's note: "total cost values" see Peacock: 59) which correspond to x_1 and x_2, and therefore $|K(x_1) - K(x_2)|$. It also follows however that there is only a finite number of points of discontinuity which have variations greater than α. The number of these points cannot be greater than quotient

$$\frac{K(x_2) - K(x_1)}{\alpha}.$$

We can ignore all points of discontinuity which have a variation smaller than α because the value of the total cost values can be changed so that they differ from the original values by less than α and the newly defined total cost function is continuous. Furthermore, where the function is continuous, we can bring about regularity through changes which are smaller than α. We have therefore established our proposition.

Average costs and average variable costs can always be easily calculated for any total cost function. Using the proposition established a moment ago, we are able to construct a marginal cost function for the general total cost function too. From now on, let us call the number of points of discontinuity, u. Two neighbouring points of discontinuity form a regularity interval. Within this interval, the formation of the marginal cost function is very clear. We now investigate how we must usefully define the marginal cost function at the points of discontinuity.

At the same time, the marginal costs associated with a velocity of production provide justification for a particular assertion about the total costs associated with this velocity of production. It follows that only the derivatives on the left side are key to the points of discontinuity in defining the marginal cost function at these points. We need the marginal cost function to determine the minimum position, the optimum position and the firm's supply. The relevant velocities of production were each derived from the abscissae of the marginal cost intersection points with the corresponding curves. In our example, several such intersection points can occur within the regularity interval. Then we can compare the results. However, the relevant velocities of production can also exist at a point of discontinuity. In formal terms, as each of the exceptional velocities of production in question exists because of an intersection point, we can define the marginal cost values at the points of discontinuity as follows.

The marginal cost values at a point of discontinuity are:

1. The left-sided derivative of the total cost function at this point.
2. All values which are greater than this derivative.

Using this definition, all laws that have been derived remain applicable, including the general total cost function for single supply. This fact can be proven more precisely using mathematical techniques. However, the evidence is complicated and does not fundamentally enrich economic thought. Therefore, we will not present it. Overall, the observations set out in this section are only significant for one reason. They must extend the application of the derived propositions as far as possible.

Finally, an analytical text is briefly provided to explain the facts.

Total cost curve $K\ (x)$ is regular and smooth when $u_n < x < u_{n+1}$, (where $n = 1, 2\ \ldots\ldots$). It follows that:

$$K\,(u_n) = \lim_{x \to u_n - 0} K\,(x).$$

As a result, $K\ (x)$ is defined for all $x > 0$. $K^*\ (x)$ and $K_{II}^*\ (x)$ also result from this. They have the same points of discontinuity as $K\ (x)$.

$K'\ (x)$ is defined for $u_n < x < u_{n+1}$ using this equation:

$$K'\ (x) = \frac{d\,K(x)}{d\,x}.$$

In addition, it follows that:

$$K'\ (u_n) \geqq \lim_{x \to u_n - 0} K'\ (x).$$

K' is then infinitely ambiguous in u_n.

These relationships and the application of the laws derived earlier are shown in the second example in Appendix D.[15]

[15] An exceptional feature occurs if there is an upper limit for the velocity of production, above which it cannot increase, perhaps for technological reasons. Then the total cost function is only defined between 0 and this upper limit. Here we regard the upper limit of the velocities of production to be one point of discontinuity for the total cost function. The other point of discontinuity then follows from the remarks made already. The brickworks factory (Translator's note: "Ziegeleifabrik" in German) provides us with an example to illustrate this fact (see Enquête-Ausschuss (Survey) I, 3rd working copy, Part 2, 2: 186).

Appendix C. Remarks About Eugen Schmalenbach's Cost Theory

§ 1. Theory of Single Production

Eugen Schmalenbach's work has a central place in German cost theory, especially his *Grundlagen der Selbstkostenrechnung und Preispolitik*.[16] This book, which was first published in 1909 as an article in *Zeitschrift für handelswissenschaftliche Forschung* (Review for Commercial Research), provides the basis for today's German cost theory. Therefore it would seem reasonable to undertake an analysis of the main propositions from this paper. At the same time, we are also presented with a theoretical application of formal cost theory, that is, an analysis of cost theories that are not based on mathematics.

The three main points of Schmalenbach's cost theory are:

I. The division of costs
II. The proportional cost
III. Cost estimation.

I.

The division of costs.

Schmalenbach distinguishes between fixed, degressive, proportional and progressive costs.

1. Fixed costs: Schmalenbach defines these as follows: "Their nature [The nature of the fixed costs: v. Stackelberg's note] ... is such that they are not influenced by a fluctuating employment level".[17] Schmalenbach uses "employment level"

[16] 5th German ed., Leipzig, 1930 (Translator's note: published in English as: *Dynamic Accounting and Price-Level Adjustments: An Economic Consequences Explanation* (1989)).

[17] Ibid.: 37.

DOI 10.1007/978-3-642-34537-1, © Springer-Verlag Berlin Heidelberg 2014

to mean "the quantity of products currently produced".[18] Here, we should probably add: "The quantity of products currently produced in the unit time." Schmalenbach's term "employment level" then equates to our term "production level" and so fixed costs are simply our constant costs K_{I}. However, the unambiguous, and in our opinion, also appropriate term, "fixed costs", is apparently not kept to describe the "dismantling of costs" and so in order to prevent any misunderstanding, we have used the term "constant costs" where the word "constant" exactly equates to the mathematical nature of this cost function.

Business managers are often heard to argue that fixed costs are also not independent of the employment level but instead "jump" or change rapidly with an increase in employment levels. This is however clearly only the case if we use a different definition to Schmalenbach for the term, "fixed costs". "Jumping costs" are then also included in fixed costs. In our opinion, this broadening of the term "fixed costs" is inappropriate. Here, we prefer to introduce the specific term "jumping costs" instead.

2. Degressive costs: Schmalenbach defines these as follows: "Degressive total costs are characterised by the fact that all costs actually increase with an increasing employment level but the gradient is smaller than the production gradient[19]". This definition is logical. Based on the numerical example set out in the "calculation of actual costs", we can interpret from this that the cost gradient relative to total costs is smaller than the gradient of the velocity of production relative to the velocity of production itself. It therefore follows that:

$$\frac{K\,(x+\Delta\,x)-K\,(x)}{K\,(x)}<\frac{\Delta\,x}{x}$$

if $\Delta\,x$ is positive and sufficiently small. It follows that:

$$\frac{K\,(x+\Delta\,x)-K\,(x)}{\Delta\,x}<\frac{K\,(x)}{x}=K^{*}\,(x)$$

or in the equilibrium:

$$K'\,(x)<K^{*}\,(x).$$

The inequality is then fulfilled if average costs decrease (see propositions III and IIIa). This is also consistent with the numerical example presented by Schmalenbach.

We can establish that our definition of degressive costs is covered by Schmalenbach's definition. It therefore follows that degressive costs are costs whose average decreases with an increasing velocity of production.

---

[18] Ibid.: 32.

[19] Ibid. (German ed.): 37.

3. Proportional costs: Schmalenbach describes these in the following way: "Where the employment level drops back to half, costs therefore decrease by half and where the quantity of goods produced doubles, costs double[20]". It generally also follows that:

$$K\ (x) = x\ .\ K\ (1)$$

and therefore

$$K'\ (x) = K\ (1).$$

We hence derive:

$$K^*\ (x) = \frac{x\ .\ K\ (1)}{x} = K\ (1) = K'\ (x)$$

as a condition for the fact that costs are "proportional".
4. Progressive costs: Similarly to degressive costs, we show that progressive costs satisfy this condition

$$K'(x) > K^*\ (x)$$

and therefore this leads to our definition of Schmalenbach's condition here too.

It is highly appropriate to introduce the terms, "degressive" and "progressive" for the purposes of using shorter terminology. Conversely, the term, "proportional costs" appears unnecessary because total costs are usually only "proportional" at one point, that is, in the optimum position (similarly, variable costs in the minimum position).

II.

The proportional cost.

Where x_1 and x_2 are two different values for velocity of production x, the corresponding proportional cost, which we shall describe using Q is:

$$Q = \frac{K\ (x_2) - K\ (x_1)}{x_2 - x_1}.$$

[20] Ibid. (German ed.): 32.

The proportional cost is therefore the differential quotient for the total cost function which corresponds to the two velocities of production x_1 and x_2. We can also abbreviate this by writing:

$$Q = Q\ (x_1,\ x_2).$$

Schmalenbach has named the proportional cost (see footnote 20) as the "basis of calculation in many cases".[21] The importance of this statement arises less from theoretical aspects (the proportional cost is just an approximate value for marginal costs. Marginal productivity, which puts forward the proposition, "marginal costs equal price", already traces its roots back to Ricardo and Thünen) than from practical aspects. Schmalenbach's contribution is that he was the first to have introduced this value as the basis of calculation in practical cost calculations. He has shown that under certain conditions so much must be produced that a proportional cost which corresponds to the value of the velocity of production achieved, is approximately equal to the market price. The more accurate this proposition is, as is clear from our remarks, the less the proportional cost differs from the marginal costs for the velocity of production achieved.

III.

The dismantling of costs.

We must take more time to discuss the "dismantling of costs" than we did for the points discussed a moment ago. There is an extraordinary amount of confusion and misunderstanding about this subject alone. Schmalenbach's remarks are not at all clear and appear to us to need interpretation. They have often been interpreted and developed too. However, we know only of a single interpretation which we can possibly describe as an entirely faultless and complete interpretation, and this can be found in the article by Dr. Kosiol, "Kostenauflösung und proportionaler Satz" (Cost Estimation and Proportional Costs) in *Zeitung für handelswissenschaftliche Forschung* (Review for Commercial Research), 1927. Later, an article by Dr. Kalischer was also published ("Der Widerspruch zwischen mathematischer und buchtechnischer Kostenauflösung" (The Contradiction between Mathematical and Accounting Cost Estimation) in *Zeitung für handelswissenschaftliche Forschung* (Review for Commercial Research), April 1929), in which the break-down of costs is interpreted correctly. However, Kalischer goes beyond giving a mere interpretation and makes his observations based on a specific case in which

[21] Ibid.: 52. On pp. 53 of the German ed., Schmalenbach, E. explains that transfer prices must also be set based on marginal costs. (Translator's note: "basis of calculation" see Peacock, A.: 32). Comparing this assertion with the observations we made in the fourth section of Chap. 3 shows that we can only agree if this good is clearly marketable under free competition.

marginal costs decrease and the law of increasing returns then applies. Schmalenbach, like Kosiol, fails to investigate this case because he simply offers his interpretation. Therefore, we cannot agree with Kalischer when he says that Kosiol could have been misled by Schmalenbach's example (Translator's note: "irreführen lassen" means "misled". v. Stackelberg cites the German ed. pp. 177). Schmalenbach's example does not illustrate "any unusual relationships" either, but instead it is just a special case, that is, one in which marginal costs increase from the start so that the firm is therefore subject to the law of diminishing returns.

We will refer to the article by Kosiol mentioned above concerning the interpretation of Schmalenbach's "breakdown of costs" because we consider Kosiol's explanations to be definitive. We can therefore move on to an analysis.

Schmalenbach breaks down costs into a proportional and a fixed "component". It would be an error however to say that the fixed component would have something to do with fixed costs as defined above. This point is not entirely clear in Schmalenbach's work. Kosiol, however, presents the only possible viewpoint. "Fixed component" would be a very unfortunate choice of name anyway. However, the fact that this "fixed component" is described as "fixed costs" in the "calculation of actual costs" in the example on page 45 (above), suggests that Schmalenbach also thought that the fixed component and fixed costs were identical. This impression is further reinforced by Schmalenbach's observations in his book, *Der Kontenrahmen*.[22] Here he seems to believe, like Maletz,[23] that there are only fixed and proportional costs, and that degressive and progressive costs are made up of the other two categories. However, we can establish that unless we are really using the term "proportional" in its mathematical sense, this thesis cannot then continue to be supported with regards to progressive costs and only with regard to degressive costs if these are linear, that is, if the total cost function is linear and degressive.

Costs are dismantled using the following method: the proportional cost is multiplied by one of its two velocities of production and the product of its total costs is then subtracted. The remainder (the same difference occurs in both cases) is described by Schmalenbach as being a fixed component of total costs. If we keep the terms we used above and let the symbol f stand for the "fixed component", it therefore follows that:

$$K\ (x_1) - x_1.\ Q\ (x_1,\ x_2) = K\ (x_2) - x_2.\ Q(x_1,\ x_2) = f = f(x_1,\ x_2).$$

There is no explicit definition of the length of the interval $(x_1,\ x_2)$ in Schmalenbach's article. However, this is because he makes the assumption that total costs are linear within this interval[24] and so the interval must be small, since

[22] Ibid.: 31 et seq. (Translator's note: published in English as: *Framework of Accounts* see Seicht (1977):15, translated in *Two Hundred Years of Accounting Research*, Seicht, G., Mattessich, R. (2008): 62), New York, Routledge.

[23] Maletz (1926): 293.

[24] Ibid.: 28. Schmalenbach, E. uses the term "standard" here, but what he probably means is "linear".

the more accurate this assumption is, the smaller the interval is, and it is therefore at its most accurate when it converges to zero. Then however Q becomes K' and we obtain the "fixed component" as a function of the velocity of production using this equation:

$$f = K\ (x) - x\ .\ K'\ (x) = f\ (x).$$

f is usually constant within even the smallest interval.[25] There is only one point at which f coincides with K_1 and that is the minimum position. Let

$$K_{\mathrm{I}} = f = K\ (x) - x\ .\ K'\ (x)$$

and it follows that

$$x\ .\ K'\ (x) = K\ (x) - K_{\mathrm{I}} = K_{\mathrm{II}}\ (x)$$

$$K'\ (x) = \frac{K_{\mathrm{II}}\ (x)}{x} = K^*_{\mathrm{II}}\ (x).$$

However, this equation is only fulfilled at the minimum position although we can determine the shape of a function which satisfies the condition that the fixed component is equal to fixed costs in terms of its overall shape. Then the above equation must be fulfilled for all values of x. The total cost function, obtained by integrating this differential equation, is linear. K' (marginal costs) is constant. The general form of a function of this kind is:

$$K\ (x) = K_{\mathrm{I}} + x\ .\ K'.$$

In addition, this is a very particular function which cannot occur in the competitive economy because it is subject to the law of constant returns.[26]

It is perhaps interesting to show the difference between "fixed costs" (=constant costs) and the "fixed component" geometrically. In Fig. C.1, both "fixed costs" and the "fixed component" are reduced on the ordinate axis. "Fixed costs" are measured using line segment $\overline{OA}$. The "fixed component" that corresponds to a velocity of production x, is always the line segment from the origin to the intersection point of the ordinate axis, and the tangent to the total cost curve at the point which corresponds to the relevant velocity of production (point $[x, K\ (x)]$).

[25] According to the initial wording of Schmalenbach's text, if we let the variable Q depend on the two employment levels x_1 and x_2, and hence let them be regarded as differential quotients, the outcome remains the same. The description would be considerably more complex.

[26] Therefore it is unacceptable to regard the total cost function of a competitive enterprise as being only approximately linear, as Lehmann suggests in his article (Lehmann 1926: 146).

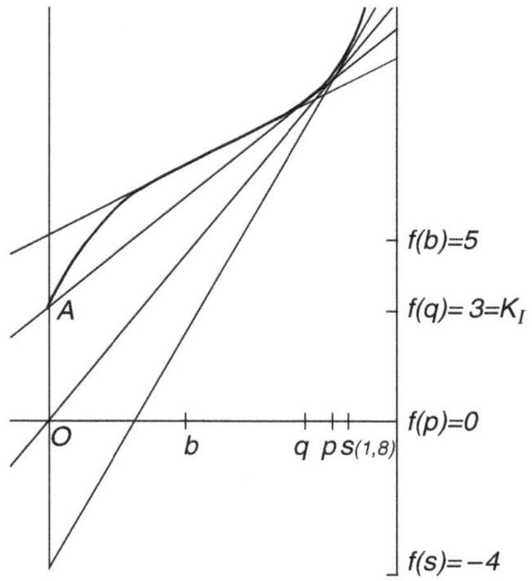

Fig. C.1 Fixed costs and the fixed component

f is positive when there is cost degression and negative when there is cost progression. We can then either use f or even $\frac{f}{K_I}$ as a measure of degression or progression. f is the loss incurred (when it is positive), or the profit (when it is negative), if velocities of production are supplied at their marginal costs (that is, in the competitive economy). However, it is impossible to easily infer the level of "fixed costs" from the level of the "fixed component".

We can summarise our results:

1. The assertion that degressive costs are the sum of a "proportional" and a "fixed" component (and similarly, the corresponding assertion for progressive costs) is just one part of the definition of the fixed component.
2. It is incorrect to claim that degressive costs are the sum of proportional costs and fixed costs (in line with the original definition of fixed costs). It is not possible to obtain an arbitrary degressive cost function in which a linear function is added to fixed costs, that is, added to either a constant function or, to one which might be step-shaped at best.
3. The term "fixed return", which is used with dismantling progressive costs, has absolutely nothing to do with fixed costs.
4. The expression, "fixed component", is inappropriate because it is impossible to see which feature is described by the adjective "fixed".

§ 2. Theory of Joint Production

We can safely assume that the contents of the section entitled, "Der Kalkulationswert bei Kuppelprodukten" (The Basis of Calculation for Joint Products)[27] is very familiar so that a word-for-word citation is unnecessary. It is now a question of producing two goods in a fixed proportion. An exceptional feature follows from the demand side being ranked according to a usage scale, whereas the costs of the velocity of production are proportional. We know that this is only possible if the enterprise has a monopoly. In this sense, we can also in fact explain the usage scale. Let us make the following assumption: if we obtain 150 Mk for 100 kg and sell 1,000 kg of product A, we therefore obtain a revenue of 1,500 Mk (Translator's note: Mk refers to the former currency of German Marks. Also, 1 kg = 2.20462 lbs). If, instead of selling 1,000 kg, we sell 1,000 kg + 1,500 kg = 2,500 kg (all of it at one price), revenue increases from 1,500 Mk to 1,500 Mk + $\frac{1500\,.\,120}{100}$ Mk = 3,300 Mk. As a result, an increase in sales of 1,500 kg leads to an increase in revenue of 1,800 Mk, that is, 120 Mk per 100 kg. A total revenue of 3,300 Mk for 2,500 kg means that this is sold for 132 Mk for 100 kg and so on. We can then derive a price function for product A from the usage scale and similarly, for product B.

If we denote the revenue of a particular quantity x of product A using $E\,(x)$, then by denoting the first quantity on the usage scale using x_1, the sum of the first and second using x_2, the sum of the first, second and third using x_3, we can then state a general formula for preparing a "usage scale" from the price function as follows:

Since revenue $E\,(x_1)$ is obtained for x_1, the quantity unit of x_1 is measured using $\frac{E(x_1)}{x_1}$. This is however just the price of x_1, and therefore $P\,(x_1)$. The increase of $E\,(x_2)$ compared to $E\,(x_1)$ is obtained by the additional quantity sold x_2-x_1. The second reason for usage therefore gives the following as a measurement of the quantity unit:

$$\frac{E(x_2) - E(x_1)}{x_2 - x_1};$$

similarly for the third increased quantity x_3-x_2

$$\frac{E(x_3) - E(x_2)}{x_3 - x_2}$$

and so on.

We can see that we have to do this with differential quotients for function $E\,(x)$. By allowing the increased quantities to pass through a zero profit, we obtain the details of the usage values of the quantity unit at an arbitrary point, that is, the value

[27] Schmalenbach, E. in the German ed.: 28 et seq.

per unit that reaches a minimum increase in a particular velocity of production. We can therefore simply replace the usage scale with the derivative of revenue as a function of the velocity of production (and indeed, this is the precise approach). In other words, we have to replace the usage scale with marginal revenue function $E'(x)$.

Where price is constant, free competition then prevails and the usage scale consists of only one usage step with an arbitrary production quantity and a usage value that is equal to the price. In the example we find in Schmalenbach, we are however dealing with a monopoly and must indeed be doing so because the cost function is assumed to be linear. Based on the fundamental proposition of profit-making production, the marginal usage value (that is to say, the marginal return) must be equal to marginal costs.

The example that Schmalenbach himself provides[28] will be examined in a fairly changed form. We are now dealing with a two-dimensional vector whose components are x and y. The following fixed proportion applies: $x{:}y = 1{:}3$. The quantity unit will be represented by the vector (25, 75) (both goods are measured in kilograms). We summarise the reason for usage of the two products A and B so that a constant (average) usage value exists for each product within a level. We hence obtain the following table which is clearly influenced by Schmalenbach's diagram, as you will probably clearly see.

	Total production		Usage quantities				Usage values per 100 kg		Usage value of the component of the vector	
Usage steps	For A	For B	For A		For B		For A	For B	(25,75)	Usage steps
1	1,000	3,000	1,000		3,000	}5,000	150	800	637.50	1
2	1,667	5,000	667	}1,500	2,000		120	800	630.00	2
3	2,500	7,500	833		2,500	}4,000	120	700	555.00	3
4	3,000	3,000	500	}1,000	1,500		100	700	550.00	4
5	3,500	10,500	500		1,500	}6,000	100	650	512.50	5
6	5,000	15,000	1,500	}3,000	4,500		70	650	505.00	6
7	6,500	19,500	1,500		4,500	}8,000	70	500	392.50	7
8	7,667	23,000	1,167	}2,000	3,500		50	500	387.50	8
9	8,500	25,500	833		2,500	}7,000	50	400	312.50	9
10	−10,000	30,000	−1,500	−1,500	4,500		-	400	300.00	10
			−10,000	**−10,000**	**30,000**	**30,000**				

The most favourable product vector can be seen from this table where marginal costs are constant and amount to 450 Marks per vector unit for every velocity of production. This includes usage steps 1–6 and is (5,000, 15,000).

When we studied Schmalenbach's calculation of the most favourable production level, we admit that we were tempted in exactly the same way he was. He also

[28] Ibid. (German ed.): 29 and 30.

ultimately calculates the marginal usage value for the combined product and compares it to the marginal cost rate (which is now simultaneously the average cost rate). By contrast, we must describe his attempt[29] at presenting a specific basis of calculation for each of the joint products as impracticable. It is not just that the calculations are error free (this would be the case if the marginal usage value and the marginal cost rate were exactly the same), but in our opinion their objective is fundamentally impracticable (based on our earlier observations[30] about cost allocation for joint products). Of course we only ought to say something against the principle, and not about the possibility of carrying out a separate calculation for the practical example, which is only useful for studying the main product as an original product and reducing the revenue of the by-product for the total costs of the main product.[31]

[29] Ibid. (German ed.): 30–31.

[30] See German ed. pp. 57, footnote 6.

[31] See Introduction to Chap. 3.

Appendix D. Remarks and Examples for Practical Analysis

I.

The first and fundamental aim of the cost calculation of an existing enterprise must be to determine the total cost function as precisely and fully as possible. This is because all other relevant functions are derived mathematically from the total cost function. The total cost function is calculated using a table. In principle, it is possible to express functions for any number of independent variables using a table. However, the difficulty of using such a table increases with the increase in the number of variables (product vector components) until it becomes immeasurably difficult. If we have 10 values for a one-dimensional function, then to achieve the same accuracy for an n-dimensional function, we need 10^n values.

The total cost function does not need to be determined for its entire defined area, but only inside the area in which production is possible. Here too, the empirically determined function values do not need to be excessively close to each other. Using interpolation, we can obtain the intermediate values.

We should note that the table values either shift as a result of two factors or, expressed mathematically, the functional relationship changes due to the change in the purchase price or the change in the production method, especially as a result of rationalisation. Business studies has developed methods to eliminate price variations, namely, using adjustment accounts which allow a purchased factor of production to be passed on to the original firm at the fixed transfer price. The cost table of price variations remains unaffected by this and the original function that applies at a particular point in time emerges from the table with a relatively simple correction. Now however, there may be changes in production methods due to both of these price variations as well as technological and organisational advances. Then, this table will mostly be useless and must be replaced by a new one.

When the total cost function has been established approximately in this way, we have achieved the most difficult part. We can already use the table to obtain all the variables we need for regulating velocities of production or for price adjustment policy via simple calculations. An additional aim is to research and simplify the

H. von Stackelberg, *Foundations of a Pure Cost Theory*,
DOI 10.1007/978-3-642-34537-1,

total cost function. This is firstly prepared so that costs are calculated as far as possible using an empirical definition. However, this means we must make sure that we separate out costs which contradict the cost principles e.g. we do not immediately separate out costs that can be allocated either according to the proportion of allocated costs to individual products or to intermediate products.

Classifying costs in this way enables useful causal relationships to be established, especially with changes to production methods. This is because it might sometimes be possible to avoid making a new version of the cost table in certain circumstances. Furthermore, it is easier to summarise simple sequences of numbers using analytical expressions and then to synthetically build up the total cost function from these partial functions. Mathematical statistics have the final word here. These methods allow relatively precise results to be obtained based on empirical data. An example of this can be seen in German ed. pp. 117, footnote 1 about refining the proportional cost.

II.

1. We must pay special attention to the question of whether it is possible to obtain an analytical expression which can serve as a general estimator for regular and continuous total cost functions from the theoretical observations made. We obtain an analytical estimate of total costs in a specific case where multiple function types are tested on empirical data and the best one is selected. The purpose of this is to obtain a formula which is both as simple as possible and as precise as possible. If we succeed in theoretically determining a "general formula" for a particular subject, it is indeed by no means certain, but probable at any rate, that the "general formula" represents the best estimate in individual cases. We would then also have to test the general formula every time, alongside other analytical expressions.

   The "general formula" problem plays a large role in many areas of statistics, such as Pareto's famous "courbe des revenues",[32] which is a two-parameter to four-parameter estimator for the income distribution in a market economy. Again, from a theoretical observation Gibrat[33] attempts to derive a general "proportional effect" formula for an entire sequence of "economic inequalities", that is, the "law of proportional effect" (loi de l'effet proportionnel), and it seems that he has been very successful.

---

[32] Pareto, V.: 299 et seq., especially pp. 305/306. (Translator's note: "courbe des revenues" means "revenue curve" in English).

[33] Gibrat, R. (1931).

Moore[34] selects estimators for an entire sequence of economic functions (demand, supply, costs and production), which he obtains as follows. For a function $y = f(x)$ which is to be determined, he equates the term $\frac{x}{y} \cdot \frac{dy}{dx}$ to a polynomial of the first, second or zero degree and obtains $f(x)$ by integrating the resulting differential equation. In a very similar way, he determines the functions of multiple variables. In addition, he also gives polynomial estimators for the function $f(x)$ itself.

2. Various results from our theory suggest approximating the regular total cost function using a polynomial of the third degree. The reasons for this are:

1. The total cost curve is monotonically increasing,[35] it has a point of inflexion[36)] for $x = b$, and goes to infinity without an asymptote.[37]
2. Marginal costs decrease to $x = b$[38] and increase monotonically for $x > b$,[39] where they increase beyond all limits (see footnote 38).
3. The marginal cost gradient is negative for $x < b$ (see footnote 39) and positive for $x > b$ (see footnote 37) and can be hence estimated satisfactorily using a straight line with a positive gradient which intersects the x-axis at point b.

We therefore derive the "general formula":

$$K(x) = A + Bx + Cx^2 + Dx^3. \tag{62}$$

We can make certain statements about (62):

(a) Using $K(0) = K_{\mathrm{I}}$, it follows that:

$$K_{\mathrm{I}} = A > 0. \tag{63}$$

(b) It is

$$K'(x) = B + 2Cx + 3Dx^2. \tag{64}$$

Since $K(x)$ increases monotonically and is the minimum of the gradient at $x = b$, it therefore follows that:

$$B = K'(0) > 0. \tag{65}$$

---

[34] Moore, H.L. (1929).

[35] See German ed. pp. 10.

[36] Gibrat, R. (1931).

[37] Proposition XVIII.

[38] See German ed. pp. 34.

[39] See German ed. pp. 25.

(c) It is

$$K''\ (x) = 2\ C + 6\ D\ x. \tag{66}$$

Since $K''\ (x)$ is a line with a positive gradient, as we mentioned earlier, it follows that:

$$D > 0. \tag{67}$$

According to the definition of b, it follows that:

$$2\ C + 6\ D\ b = 0. \tag{68}$$

From (68), follows:

$$C < 0 \tag{69}$$

and

$$b = -\frac{C}{3\ D}. \tag{70}$$

(d) It also follows that:

$$K^* = \frac{A}{x} + B + C\ x + D\ x^2 \tag{71}$$

and

$$K^*_{\mathrm{II}} = B + C\ x + D\ x^2. \tag{72}$$

Using (64) and (72) and according to proposition IV, it follows that:

$$B + 2\ C\ q + 3\ D\ q^2 = B + C\ q + D\ q^2$$

or

$$q\ (C + 2\ D\ q) = 0.$$

Using[40] $q \neq 0$, we obtain:

$$q = -\frac{C}{2D} \tag{73}$$

[40] Proposition V would not apply for $q = 0$. Using (69), it is $K''\ (o) = 2\ C < 0$.

from which, using (70), it follows that:

$$2q = 3\,b \tag{74}$$

(e) According to proposition I, it follows that:

$$B + 2\,C\,p + 3\,D\,p^2 = \frac{A}{p} + B + C\,p + D\,p^2$$

or

$$A - C\,p^2 - 2\,D\,p^3 = 0 \tag{75}$$

whereby using (73), it also follows that:

$$p^3 - p^2\,q = \frac{A}{2\,D}. \tag{76}$$

(f) From the fundamental proposition of the profit-making principle, it follows that:

$$E'\ (s) = B + 2\,C\,s + 3\,D\,s^2. \tag{77}$$

For free competition, it follows that:

$$P = B + 2\,C\,s + 3\,D\,s^2 \tag{78}$$

whereby following some remodelling:

$$s = \frac{1}{3\,D}\left(-C + \sqrt{3\,D\,(P - B) + C^2}\right). \tag{79}$$

The negative value of the root is not possible using the second maximum condition ($K''\ (s) > 0$).

Using (70), it also follows that:

$$s = b + \sqrt{\frac{P - B}{3\,D} + b^2}. \tag{80}$$

III.

Where the total cost function for an enterprise is calculated as either a table, a curve or an analytical expression (Translator's note: "analytical expression" see Peacock: 46), this is used to regulate production. Below are two examples. We firstly assume a regular and smooth, as well as a regular total cost function, given by an analytical expression. We secondly assume a regular total cost function with multiple points of discontinuity, given by a curve.

1. Example:

   (a) The total cost curve in Fig. 2.2 in Chap. 2 is based on this equation[41]:

$$K\ (x) = 24 + 4.6\ x - 0.6\ x^2 + \frac{1}{30}\ x^3. \tag{81}$$

From this, the following analytical expressions follow for the other curves:

$$K'\ (x) = 4.6 - 1.2\ x + 0.1\ x^2$$
$$K''\ (x) = -1.2 + 0.2\ x$$
$$K^*\ (x) = \frac{24}{x} + 4.6 - 0.6\ x + \frac{1}{30}\ x^2$$
$$K^*_{\text{II}}\ (x) = 4.6 - 0.6\ x + \frac{1}{30}\ x^2.$$

For b and using (70), we obtain: $b = 6$.
For the minimum position and using (73) or (74), it follows that: $q = 9$.
The conditional equation applies for the optimum position, using (76):

$$p^3 - 9\ p^2 = 360.$$

This equation is fulfilled for $p \approx 11.65$. Proposition II is also fulfilled:

$$K''\ (p) = -1.2 + 0.1\ .\ 11.65 = 1.13 > 0.$$

For the enterprise's supply and using (80), after calculation the following applies:

$$s = 6 + \sqrt{10\ (P - 1)}.$$

[41] The ordinates of the total cost curve are drawn with a scale four times smaller than the ordinates of the other curves.

To determine the supply function more precisely, we observe that the supplied quantity, if any at all is produced, can never be smaller than the minimum position and the supply function is only defined using the analytical expression above for those prices which are not lower than marginal costs in the minimum position. For all prices which are lower, the quantity supplied is equal to zero. In the minimum position, marginal costs amount to:

$$K'\ (9) = 4.6 - 1.2\ .\ 9 + 0.1\ .\ 9^2 = 1.9.$$

For the analytical expression above, we therefore derive these two inequalities:

$$s > 9;\ P > 1.9.$$

From the first inequality, it follows that the positive value of the above root must always be set. The second inequality, which comes from the first, expresses the fact that using the analytical expression above, the supply function is only defined for $P > 1.9$. s vanishes in an identical way for all $P < 1.9$. It now follows that $s = 0$.

As a result, we have determined the supply function as follows:

$$\begin{aligned} &\text{for}\quad P > 1.9\quad \text{it follows that}:\quad s = 6 + \sqrt{10\ (P - 1)};\\ &\text{for}\quad P < 1.9\quad \text{it follows that}:\quad s = 0. \end{aligned}$$

6. The optimum price is equal to the marginal costs in the optimum position. We derive:

$$K'\ (p) = 4.6 - 1.2\ .\ 11.65 + 0.1\ .\ 11.65^2 \approx 4.2.$$

The optimum price is therefore 4.2.

The enterprise can only achieve a velocity of production of between 9 and 11.65 at a price of between 1.9 and 4.2. Here, it obtains a gross profit which does not completely cover constant costs. It therefore incurs a loss because it has constant costs of 24 and a gross profit of less than 24.

Where the price is greater than 4.2, the enterprise then achieves a velocity of production greater than 11.65. Its gross profit is greater than 24 and it obtains an additional profit.

7. Where the price is 3.5, for example, the enterprise's supply is then:

$$s = 6 + \sqrt{10\ (3.5 - 1)} = 11.$$

Total costs are now:

$$K\ (11) = 24 + 4.6\ .\ 11 - 0.6\ .\ 11^2 + \frac{11^3}{30} \approx 46.37.$$

Variable costs are: $K\ (7) - K_{\mathrm{I}} = 22.37$.
Revenue: 11 . 3.5 = 38.5.
Gross profit: 38.5 – 22.37 = 16.13.
Net loss: 24 – 16.13 = 7.87.

8. Where the price is 5.9, the enterprise's supply is:

$$s = 6 + \sqrt{10(5.9 - 1)} = 13.$$

Total costs are now:

$$K(13) = 24 + 4.6\,.\ 13 - 0.6\,.\,13^2 + \frac{13^3}{30} \approx = 55.63.$$

Variable costs are: $K\ (13) - K_{\mathrm{I}} = 31.63$.
Revenue: 13 . 5.9 = 76.7.
Gross profit: 76.7 – 31.63 = 45.07.
Net profit: 45.07 – 24 = 21.07.

9. If the enterprise were only to achieve its optimum position, it would then always be in a worse position if the price were to differ from 4.2.
With a price of 3.5, the net loss would amount to:

$$[K^*(p) - P]\,.\,p = (4.2 - 3.5)\,.\,11.65 = 8.15.$$

It is then 0.28 more than if a velocity of production of 11 had been achieved.
With a price of 5.9, the net profit would amount to:

$$[P - K^*(p)].p = (5.9 - 4.2)\,.\,11.65 = 19.80.$$

It is then 1.2 less than if a velocity of production of 13 had been achieved.
The fact that the divergences are relatively small (3.6 % of the current most favourable amount in the first case and 5.7 % in the second) is explained by the fact that the elasticity of our supply curve is fairly small. For example, it is 0.319 in the optimum position.

2. Example (Fig. D.1.).

Marginal cost function $K'\ (x)$, average cost function $K^*(x)$ and average variable cost function $K_{\mathrm{II}}^*\ (x)$ have been marked on a general total cost function $K\ (x)$. The total cost function has been drawn to a scale five times smaller than the other functions.

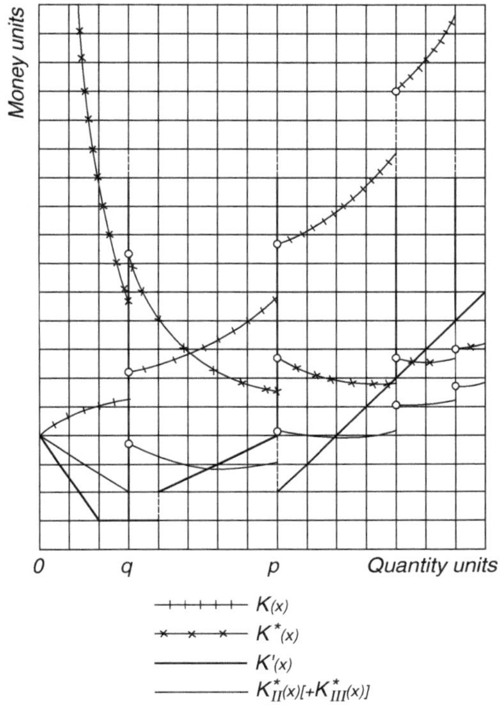

Fig. D.1 The most favourable velocity of production

The small circles (o) mean that the ordinates of the relevant points are not a function value of functions $K(x)$, $K^{*}(x)$ or $K^{*}_{\mathrm{II}}(x)$.

We can extract the following facts from the graph.

1. The minimum velocity of production is 3. Here, average variable costs reach their minimum position using $K^{*}_{\mathrm{II}}(3) = 2$. Formally, an intersection point with the marginal cost curve exists.
2. The optimum position is where $x = 8$. Here, average costs reach their minimum position using $K^{*} = 5.5$. The laws governing costs are formally applicable here too.
3. Where the price is $P = 5$, the velocities of production 3, 8 and 11 have marginal costs which are equal to the price and because the price is less than the optimum price, $x = 11$ is ruled out. Therefore we must compare velocities of production 3 and 8 with each other.

$$G(3) = 5 \,.\, 3 - K(3) = 15 - 26 = -11$$
$$G(8) = 5 \,.\, 8 - K(8) = 40 - 44 = -4.$$

$x = 8$ is the most favourable velocity of production. It therefore follows that $s(5) = 8$.

We can easily show that $x = 11$ is not really a possibility as follows:

$$G(11) = 5 \,.\, 11 - K(11) = 55 - 63.8 = -8.8.$$

Velocity of production 11 is therefore less favourable than 8.

4. Where the price is $P = 7$, then velocities of production 3, 12 and 13 are initially possibilities. Since 7 is greater than the optimum price of 5.5, this then rules 3 out. We can therefore compare G (8), G (12) and G (13). It then follows that:

$$G\,(8) = 7 \,.\, 8 - K(8) = 56 - 44 = 12$$
$$G\,(12) = 7 \,.\, 12 - K(12) = 84 - 69.5 = 14.5$$
$$G\,(13) = 7 \,.\, 13 - K(13) = 91 - 85 = 6.$$

The most favourable velocity of production here is 12. We therefore derive:

$$s(7) = 12.$$

References

Dalton, H. (1920). The measurement of the inequality of incomes. *The Economic Journal, 30*, 348–461.

Gibrat, R. (1931). *Les inégalités économiques* [published in English as: On economic inequalities]. Paris: Librairie du Recueil Sirey.

Lehmann, M.R. (1926). "Grundsätzliche Bemerkungen zur Frage der Abhängigkeit der Kosten vom Beschäftigungsgrad" (Fundamental Observations about the Question of Cost Dependency) (in *Betriebswirtschaftliche Rundschau* (Managerial Review), 146).

Maletz, (1926). *Zeitung für handelswissenschaftliche Forschung* (Review for Commercial Research): 293.

Marshall, A. (1905). *Handbuch der Volkswirtschaftslehre* [published in English as: *Principles of economics*, 1890] (vol. 2, Book VI, Chapter XII: 685) (Translator's note: Principles of Economics was Translated into German as Handbuch der Volkswirtschaftslehre by Hugo Ephraim and Arthur Salz, Stuttghart-Berlin).

Moore, H. L. (1929). *Synthetic economics*. New York: Macmillan.

Pareto, V. (1906). *Cours d'économie politique* 2nd edition (published in English as: *Manual of Political Economy*), II: 299 et seq.

Pareto, V. (1927). *Manuel d'économie politique* (published in English as: *Manual of Political Economy*). Appendix, No. 77 et seq.

www.ingramcontent.com/pod-product-compliance
Lightning Source LLC
LaVergne TN
LVHW012323171026
850499LV00003B/83
9783642345364